PUBLISHERS' NOTE

The series of monographs in which this title appears was introduced by the publishers in 1957, under the General Editorship of Dr Maurice G. Kendall. Since that date, more than twenty volumes have been issued, and in 1966 the Editorship passed to Alan Stuart, D.Sc.(Econ.), Professor of Statistics, University of London.

The Series fills the need for a form of publication at moderate cost which will make accessible to a group of readers specialized studies in statistics or courses on particular statistical topics. Often, a monograph on some newly developed field would be very useful, but the subject has not reached the stage where a comprehensive treatment is possible. Considerable attention has been given to the problem of producing these books speedily and economically.

It is intended that in future the Series will include works on applications of statistics in special fields of interest, as well as theoretical studies. The publishers will be interested in approaches from any authors who have work of importance suitable for the Series.

CHARLES GRIFFIN & CO. LTD.

GRIFFIN'S STATISTICAL MONOGRAPHS AND COURSES

No. 1:	The analysis of multiple time-series	M. H. QUENOUILLE
No. 2:	A course in multivariate analysis	M. G. KENDALL
No. 3:	The fundamentals of statistical reasoning	M. H. QUENOUILLE
No. 4:	Basic ideas of scientific sampling	A. STUART
No. 5:	Characteristic functions*	E. LUKACS
No. 6:	An introduction to infinitely many variates	E. A. ROBINSON
No. 7:	Mathematical methods in the theory of queueing	A. Y. KHINTCHINE
No. 8:	A course in the geometry of n dimensions	M. G. KENDALL
No. 9:	Random wavelets and cybernetic systems	E. A. ROBINSON
No. 10:	Geometrical probability	M. G. KENDALL and P. A. P. MORAN
No. 11:	An introduction to symbolic programming	P. WEGNER
No. 12:	The method of paired comparisons	H. A. DAVID
No. 13:	Statistical assessment of the life characteristic: a bibliographic guide	W. R. BUCKLAND
No. 14:	Applications of characteristic functions	E. LUKACS and R. G. LAHA
No. 15:	Elements of linear programming with economic applications	R. C. GEARY and M. D. MCCARTHY
No. 16:	Inequalities on distribution functions	H. J. GODWIN
No. 17:	Green's function methods in probability theory	J. KEILSON
No. 18:	The analysis of variance	A. HUITSON
No. 19:	The linear hypothesis: a general theory	G. A. F. SEBER
No. 20:	Econometric techniques and problems	C. E. V. LESER
No. 21:	Stochastically dependent equations: an introductory text for econometricians	P. R. FISK
No. 22:	Patterns and configurations in finite spaces	S. VAJDA
No. 23:	The mathematics of experimental design: incomplete block designs and Latin squares	S. VAJDA
No. 24:	Cumulative sum tests: theory and practice	C. S. VAN DOBBEN DE BRUYN
No. 25:	Statistical models and their experimental application	P. OTTESTAD
No. 26:	Statistical tolerance regions: classical and Bayesian	I. GUTTMAN
No. 27:	Families of bivariate distributions	K. V. MARDIA
No. 28:	Generalized inverse matrices with applications to statistics	R. M. PRINGLE and A. A. RAYNER
No. 29:	The generation of random variates	T. G. NEWMAN and P. L. ODELL

*Now published independently of the Series.

For a list of other statistical and mathematical books see back cover.

THE
ANALYSIS OF VARIANCE
A BASIC COURSE

ALAN HUITSON B.Sc. Ph.D
Fellow of the Royal Statistical Society . Associate
of the Institute of Statisticians . Member of the
Biometric Society . Consultant Statistician IBM

BEING NUMBER EIGHTEEN OF
**GRIFFIN'S STATISTICAL
MONOGRAPHS & COURSES**
EDITED BY
ALAN STUART, D.Sc.(Econ.)

Second impression

GRIFFIN   LONDON

Copyright © 1971
CHARLES GRIFFIN & COMPANY LIMITED
42 DRURY LANE, LONDON, W.C.2

First published . . . 1966
Second impression . . 1971

Demy Octavo, viii + 83 pages
ISBN: 0 85264 174 5

PRINTED IN GREAT BRITAIN BY
LATIMER TREND & CO LTD, WHITSTABLE

PREFACE

The subject of analysis of variance, although now of a respectable age, is still one which presents numerous difficulties to the student. Many students end their training by regarding it as a slice of algebra to be carried out mechanically, without any understanding of the practical implications of what they are doing. While a facility in algebraic manipulation is desirable, a full understanding of the assumptions involved is necessary for the successful application of the technique. The idea for starting this monograph arose from teaching the need to differentiate between the use of the interaction mean square and the error mean square when carrying out a variance ratio test. Existing books are particularly weak on this topic, and yet it is from this single point that the unity of the subject can be established.

In this volume, I have tried to present a logical development, covering both the assumptions made and the algebraic derivations of the formulae. Only where repetition of the algebra would have become tedious has the work been left as an exercise to the reader. The first three chapters cover the conventional analysis of variance. The next four chapters show how the methods which have been developed can be applied to special types of experiments. Finally, Chapter 8 discusses some further points which arise occasionally in this field. All the theory of the subject necessary for a university degree course or a Dip. Tech. course has been included.

I would like to record my gratitude to my wife for her meticulous typing of the drafts and also to the publishers for their very careful setting of some difficult manuscript.

August 1965   A.H.

PUBLISHERS' NOTE TO SECOND IMPRESSION

The continued demand for this Monograph necessitates a reprint, and the opportunity has been taken to make some minor amendments and additions to the text.

November 1970   CHARLES GRIFFIN & CO. LTD.

CONTENTS

Chapter		Page
1	ONE-WAY CLASSIFICATION	
	1.1 Introduction	1
	1.2 Types of factors	1
	1.3 One-way classification with equal numbers of observations	3
	1.4 Unequal numbers of observations	6
	1.5 Low F-ratios	7
	1.6 Testing the overall mean	8
	1.7 Computational note	9
	1.8 Example	9
2	TWO-WAY CLASSIFICATIONS	
	2.1 Introduction	11
	2.2 Two crossed classifications with replication	13
	2.3 Two crossed classifications with no replication	17
	2.4 Two crossed classifications with no interaction	18
	2.5 Two nested classifications	19
	2.6 Missing data	21
	2.7 Computational note	22
	2.8 Example 1 – Crossed classification	22
	2.9 Example 2 – Nested classification	24
3	GREATER NUMBERS OF CLASSIFICATIONS	
	3.1 Introduction	26
	3.2 Crossed classifications with replications	26
	3.3 Nested classifications	30
	3.4 Analyses involving nested and crossed classifications	32
	3.5 Computational note	34
	3.6 Transformation of data	35
4	APPLICATION TO REGRESSION ANALYSIS	
	4.1 Introduction	37
	4.2 Linear regression on one variable	37
	4.3 Linear regression with more than one variable	39
	4.4 Example	41

Chapter	Page
5 SPECIAL TYPES OF ANALYSIS	
5.1 Introduction	43
5.2 Randomised blocks	43
5.3 Latin squares	46
5.4 Balanced incomplete blocks	48
5.5 Split plot designs	49
5.6 Examples	51
6 ANALYSIS OF COVARIANCE	
6.1 Introduction	56
6.2 Mathematical derivation for one-way classification	56
6.3 Two-way classification	58
6.4 Example	61
7 TWO-LEVEL EXPERIMENTS	
7.1 Introduction	64
7.2 Notation	64
7.3 Confounding	68
7.4 Fractional replication	72
7.5 Factors having three or four levels	74
8 FURTHER DISCUSSION	
8.1 Introduction	76
8.2 Models	76
8.3 Linear combinations of variances	78
8.4 Suggestions for further reading	79
References	80
Index	82

CHAPTER 1

ONE-WAY CLASSIFICATION

1.1 Introduction

The variation between physical observations has been the concern of statistics for many years. It received attention from Laplace and Gauss and is still being discussed in the current journals. Throughout this book we shall be concerned with measurements which cannot be reproduced exactly. This failure to repeat an observation accurately is connected in general with the necessity for taking them under different conditions. The readings may be made by different observers or at different instants of time or according to some other scheme under which different conditions operate.

In any experiment, there are always a large number of external conditions over which the experimenter has either no control or which are too difficult or expensive to control. These may take many forms, e.g. temperature, atmospheric pressure, gravitational force, concentration of a solvent, etc. Of these, a large proportion will not affect the measurement to be made. Nevertheless, unless the experimenter is very fortunate, some of these uncontrolled conditions will affect the results of the experiment. Such external conditions are usually called *factors*. We shall be concerned with factors whose effects are large and which we can examine. It is the purpose of this book to show how, by suitable choice of experiments, these factors may be investigated. In fact, an experimenter may deliberately allow a controllable factor to vary in order to investigate its effect on his experiment. We shall show how the resultant variation of a number of experiments may be resolved into component variations due to the different factors. Such a technique was originally given the name of Analysis of Variance by R.A. Fisher (1938) who defined it as "separation of the variance ascribable to one group of causes from the variance ascribable to other groups".

1.2 Types of factors

Factors may be of two kinds, those whose effects are systematic and those whose effects are random. An example of the first kind of factor would be the effect of various treatments, while the second kind could arise from the lack of homogeneity of some raw material. In one case, the experimenter is working with a systematically chosen set of treatments which are of particular interest to him, and these treatments will have fixed or systematic effects on the experiment. In the second case, when the raw material has been divided into experimental units, the effect of each unit on the experiment is not under the control of the experimenter. He has made a random selection from all the possible kinds of experimental units and hence this raw material is said to have a random effect. Sometimes it is difficult to decide whether an effect is to be regarded as random or systematic. The answer to this is usually obvious when we know the manner in which the individuals were selected. For example, suppose we are dealing with factories or works, such as coke-oven plants of which there are relatively few. If we take all the plants in a district into our experiment their effect will be systematic, and the results of our experiment are only applicable to those particular plants used in the experiment. However, if a random sample of all the plants in a given region were taken for the experiment, then their effect is random and the experimental results are applicable by induction to all plants in the region from which the sample was selected. Thus, in general, random effects correspond to random selection of individuals.

The results of an experiment carry more weight when randomisation is applied at every level of sampling since the results can be applied to the whole population from which the sample has been obtained. But often it is not possible to carry out a complete randomisation without greatly increasing the cost of the experiment, so it then becomes necessary to take a particular sample in which the factor will be considered to have a systematic effect. Whether the factors of an experiment are random or systematic, it is hazardous to apply results outside the region of the experiment. For example, that one variety of barley produces, on an average, three stones per acre more than another does not necessarily mean much to an English farmer if the experiment was carried out in Ireland, even though the selection of the plots in Ireland may have been entirely random.

Some factors, such as treatments, can only have a systematic effect as the treatments which are compared experimentally are never chosen at random, but are the ones in which the scientist happens to be particularly interested. There is no disadvantage in not being able to randomise here, as we only wish to apply our results to the treatments which are being experimentally compared.

It will be convenient to consider experiments of three kinds: (a) those in which all the factors have effects which are systematic,

(b) those where all the factors have random effects, and (c) those where there is a mixture of factors which have random and systematic effects. (a) and (b) correspond to Models I and II respectively of Eisenhart (1947). We shall call (c) the Mixed Model. Plackett (1960) has reviewed the literature concerning the various models which have been assumed by other authors.

It may be noted here that a general analysis covering Models I and II and the Mixed Model could be carried out by allowing each factor to take only a finite number of possible levels and selecting a certain proportion at random for inclusion in the experiment. As described, the factor would be random except for the case where the proportion of levels selected was equal to unity, and then the factor would have a systematic effect. Such a method of analysis was adopted by Bennett and Franklin (1954). However, it is not possible to proceed very far along these general lines, and in order to make further progress, two special cases must be considered separately: the case where the number of possible levels of the factor is infinite and the case where all possible levels are used in the experiment, i.e. a systematic factor. Thus, it will be assumed throughout this monograph that the factors having random effects mentioned in Model II and the Mixed Model are such that the number of possible levels is large enough to be considered infinite. If this restriction is not observed, very few of the tests which are described here are applicable. By analysing each experiment under the three models described instead of the general analysis, it is hoped to stress the fact that the appropriate tests in a particular case depend on the experimental conditions.

1.3   One-way classification with equal numbers of observations

Consider a single factor which has taken p different levels and suppose that n observations have been made at each level, giving $N = np$ observations in all. Let the results be represented by x_{ij} ($i = 1...p, j = 1...n$). We shall assume that, for a given level, the n observations have a mean which is a combination of an overall mean plus a variation from this overall mean due to the level chosen.

Hence we may write

$$x_{ij} = \mu + F_i + \epsilon_{ij} \qquad (1)$$

where μ is the overall mean, F_i is the effect due to the ith level of the factor and ϵ_{ij} is the variation of the results within a particular factor level. The term ϵ_{ij} takes into account all those factors which have not been controlled.

We shall assume that the observations at a fixed factor level are Normally distributed about the mean value $\mu + F_i$ with a common variance σ^2. Using the notation that a dot in place of a suffix means that that particular suffix has been averaged out over the appropriate observations, we may write

$$x_{ij} - x_{..} = (x_{i.} - x_{..}) + (x_{ij} - x_{i.})$$

Squaring both sides and summing over i and j we have

$$\sum\nolimits_{ij}(x_{ij} - x_{..})^2 = \sum\nolimits_{ij}(x_{i.} - x_{..})^2 + \sum\nolimits_{ij}(x_{ij} - x_{i.})^2$$

The cross-product term disappears since the sum of the deviations of any group of observations from their mean is zero.

$$\sum\nolimits_{ij}(x_{ij} - x_{..})^2 = n\sum\nolimits_{i}(x_{i.} - x_{..})^2 + \sum\nolimits_{ij}(x_{ij} - x_{i.})^2 \qquad (2)$$

or briefly $S = S_1 + S_2$.

Now S_2 is computed from the deviations of the N observations from the p sample means and hence has $N - p = p(n - 1)$ degrees of freedom. Similarly, S_1 is computed from the deviations of the p independent class means from the overall mean and hence has $(p - 1)$ degrees of freedom. S is, of course, based on $N - 1$ degrees of freedom.

These statements can be summarized in the analysis of variance table below, both the sums of squares and the degrees of freedom totalling correctly.

Source of estimate	Sums of squares	Degrees of freedom	Mean squares
Between levels	$S_1 = n\sum\nolimits_{i}(x_{i.} - x_{..})^2$	$p - 1$	$M_1 = \dfrac{S_1}{(p-1)}$
Within levels	$S_2 = \sum\nolimits_{ij}(x_{ij} - x_{i.})^2$	$N - p$	$M_2 = \dfrac{S_2}{(N-p)}$
Total	$S = \sum\nolimits_{ij}(x_{ij} - x_{..})^2$	$N - 1$	

Substituting from (1) into (2), the expected values we get are

$$S_1 = n\sum\nolimits_{i}(F_i + \epsilon_{i.} - F_{.} - \epsilon_{..})^2 = n\sum\nolimits_{i}(F_i - F_{.})^2 + n\sum\nolimits_{i}(\epsilon_{i.} - \epsilon_{..})^2$$

$$S_2 = \sum\nolimits_{ij}(\epsilon_{ij} - \epsilon_{i.})^2$$

The cross-product term again vanishes since the ϵ's and F's are assumed independent.

Hence S_2 is an unbiased estimate of $p(n-1)\sigma^2$, so that M_2 is an unbiased estimate of σ^2, the error variance within the levels.

If we now restrict ourselves to Model I, then the F's are fixed quantities depending on the particular levels chosen for the experiment, and S_1 is an unbiased estimate of $n\Sigma_i(F_i - F_.)^2 + n(p-1)\sigma^2/n$ since the variance of $\epsilon_{i.}$ is σ^2/n. Hence M_1 is an unbiased estimate of
$$\frac{n}{(p-1)}\Sigma_i(F_i - F_.)^2 + \sigma^2.$$

Under the hypothesis that all the levels have the same effect, i.e. all the F_i are equal, the quantities M_1 and M_2 will both be unbiased estimates of σ^2. Thus the hypothesis may be tested by computing the ratio M_1/M_2, and comparing it with the values in the tables of the F-distribution with $(p-1)$ and $(N-p)$ degrees. If the calculated value is greater than the Pth percentage point we would conclude that the hypothesis is false at the significance level P.

Having tested the hypothesis that $F_1 = F_2 = \ldots = F_p = F_.$, it is often desirable to test the significance of differences between individual levels. Suppose we wish to test whether there is a difference between the means of the levels f and g. Consider the two means $x_{f.}$ and $x_{g.}$ and define
$$x' = \frac{1}{2}(x_{f.} + x_{g.})$$

Then $S_3 = n(x_{f.} - x')^2 + n(x_{g.} - x')^2$ will have one degree of freedom as it is computed from the deviations of the means of the two independent levels from their overall mean. S_3 may be rewritten as

$$\frac{n}{2}(x_{f.} - x_{g.})^2 \tag{3}$$

Substituting into (3) from (1), S_3 becomes

$$\frac{n}{2}(F_f + \epsilon_{f.} - F_g - \epsilon_{g.})^2$$
$$= \frac{n}{2}(F_f - F_g)^2 + \frac{n}{2}(\epsilon_{f.} - \epsilon_{g.})^2 + n(F_f - F_g)(\epsilon_{f.} - \epsilon_{g.})$$

If we now set up the hypothesis that $F_f = F_g$ then the quantity S_3 will also be an unbiased estimate of σ^2 since the variance of $\epsilon_{f.}$ and the variance of $\epsilon_{g.}$ are both σ^2/n. Hence we can test the hypothesis by computing the ratio S_3/M_2 and comparing with the F-distribution with 1 and $N-p$ degrees of freedom. Again, if the calculated value is greater than the Pth percentage point we would conclude that the hypothesis is false at the significance level P.

Let us now consider this simple example under Model II. The F's are now a random sample from some specified population, which will be assumed to have a normal distribution. Note that the population from which the levels have been selected will always be known and should be specified in the conclusions of the analysis. S_1 is now an unbiased estimate of $n(p-1)\sigma_F^2 + (p-1)\sigma^2$ where σ_F^2 is the variance of the population of effects from which the levels are drawn. Hence M_1 is an unbiased estimate of $n\sigma_F^2 + \sigma^2$

We may again test the hypothesis that all the levels have the same effect by comparing the ratio M_1/M_2 with the Pth percentage point of the F-distribution with $p-1$ and $N-p$ degrees of freedom, because under this hypothesis $\sigma_F^2 = 0$.

If the test is significant, an estimate of σ_F^2 can be made by computing

$$\frac{1}{n}(M_1 - M_2) \tag{4}$$

A test for the comparison of individual levels can be carried out as under Model I.

1.4 Unequal numbers of observations

It may happen, due to a variety of reasons, that it is impossible to collect an equal number of observations at each level of the factor. Part of the data may have been lost, or certain of the levels, which are important for some other reason, may have been emphasised by taking more observations at these levels. Equal numbers of observations at each level are desirable because of the simplicity of organising the experiment, but if more data are available at some levels than at others, we must take them all into the analysis. Suppose the factor has p different levels and that n_i ($i = 1...p$) observations have been made at each level, giving $N = \Sigma_i n_i$ observations in all.

A similar analysis can be followed through as far as the table, except that S_1 becomes $\Sigma_i n_i (x_i. - x..)^2$. Under Model I the test that all the F's are equal is carried out as above. The test of significance of differences between individual levels must be modified by defining

$$x' = \frac{n_f x_{f.} + n_g x_{g.}}{n_f + n_g}$$

and
$$S_3 = n_f(x_f. - x')^2 + n_g(x_g. - x')^2$$
$$= \frac{n_f n_g}{n_f + n_g}(x_f. - x_g.)^2$$

S_3/M_2 can be again referred to the F-distribution with 1 and $N-p$ degrees of freedom.

Under Model II, S_1 becomes an unbiased estimate of

$$\left\{\frac{N^2 - \Sigma_i n_i^2}{N}\right\}\sigma_F^2 + (p-1)\sigma^2$$

The test that all the F's are equal is carried out as before, but the estimate of σ_F^2 is now

$$\left\{\frac{N(p-1)}{N^2 - \Sigma_i n_i^2}\right\}(M_1 - M_2)$$

1.5 Low F-ratios

The significance tests that we have carried out on the one-way classification have all involved comparison with the F-distribution and have all been arranged in such a way that, if the hypothesis is incorrect, the statistic will tend to be greater than one. Thus we have used some percentage point in the upper tail of the F-distribution to give us a criterion for rejecting the hypothesis. It is necessary at this stage to consider what interpretation we can make when the appropriate variance ratio gives a value on the lower tail of the variance ratio. We can anticipate getting a value of the variance ratio which is less than one, in at least 50 per cent of the cases in which the hypothesis is true. If the ratio M_1/M_2 is fractional, then, under Model II, the estimate (4) of σ_F^2 will be negative. This negative value may be interpreted as sampling fluctuations about an average value of zero.

If any of the variance ratios are significantly low at our chosen probability levels, then we must suspect the model we have assumed to describe the data and in particular the assumed randomness of the residual effects ϵ_{ij}. For example, if the tests are not carried out in a random manner, then any factor which is not controlled may have some varying effect on the sequence of experiments. This could increase the within-levels variance but leave the between-levels unchanged so that the variance ratio would be reduced. Thus, if a significantly low variance-ratio is obtained, then it is probable that an important factor which was not controlled has not been randomised during the experiment and much of the value of the experimental results has been lost.

1.6 Testing the overall mean

The overall mean of all the experiments will give an estimate of μ. Occasionally, it is necessary to test the hypothesis that the overall mean μ is equal to some constant μ_0. This can best be done by considering the quantity $S_4 = N(x_{..} - \mu_0)^2$ which has one degree of freedom. Substituting from (1) we get

$$S_4 = N(\mu - \mu_0 + F_{.} + \epsilon_{..})^2$$
$$= N(\mu - \mu_0)^2 + NF_{.}^2 + N\epsilon_{..}^2 + \text{product terms}.$$

Now the expected value of the product terms is zero, and the expected value of $\epsilon_{..}^2$ is σ^2/N. Under Model I, the expected value of $F_{.}$ is zero, so that the expected value of S_4 is

$$N(\mu - \mu_0)^2 + \sigma^2$$

Hence we may test our hypothesis by computing the ratio S_4/M_2 and comparing with the F-distribution with 1 and $N-p$ degrees of freedom.

Under Model II, the expected value of $F_{.}^2$ is σ_F^2/p if the population of levels is infinite or more generally $(1-z)\sigma_F^2/p$ if the population of levels is finite and z denotes the proportion which have been sampled. Hence the expected value of S_4 is now $N(\mu-\mu_0)^2 + N(1-z)\sigma_F^2/p + \sigma^2$.

When the population of levels is infinite, or is very large in proportion to the number considered in the experiment, then z is equal to zero and the expected value of S_4 becomes $N(\mu-\mu_0)^2 + n\sigma_F^2 + \sigma^2$ and we can test our hypothesis by computing the ratio S_4/M_1 and comparing with the F-distribution with 1 and $p-1$ degrees of freedom.

This is the first example in which Models I and II have led to different significance tests, although we have seen how the interpretation of the analysis depends on the Model which has been assumed. To test the overall mean under Model I we must compare S_4 with the "within levels mean square", but it must be compared with the "between levels mean square" if Model II is appropriate and if the population of levels is large. Thus, in this relatively simple case of one-way classification, it is necessary to decide on the appropriate model, not only for interpretation of the results, but also in order that the appropriate test may be used if it is desired to test the overall mean.

1.7 Computational note

The forms of the sums of squares given in the analysis of variance table do not represent the best forms for computation. By completing the squares we obtain

$$S = \sum_{ij}(x_{ij} - x_{..})^2 = \sum_{ij} x_{ij}^2 - \left(\sum_{ij} x_{ij}\right)^2 / N$$

$$S_1 = n\sum_i (x_{i.} - x_{..})^2 = n\sum_i x_{i.}^2 - \left(\sum_{ij} x_{ij}\right)^2 / N$$

$$= \sum_i \left(\sum_j x_{ij}\right)^2 / n - \left(\sum_{ij} x_{ij}\right)^2 / N$$

and

$$S_2 = \sum_{ij}(x_{ij} - x_{i.})^2 = \sum_{ij} x_{ij}^2 - \sum_i \left(\sum_j x_{ij}\right)^2 / n$$

Hence the calculations can be made as follows:

(1) Sum the observations for each level to form $(\sum_j x_{ij})$ for all j, and obtain the grand total $\sum_{ij} x_{ij}$
(2) Form the sum of the squares of the individual values $\sum_{ij} x_{ij}^2$
(3) Form the sum of the squares of the totals for each level $\sum_i (\sum_j x_{ij})^2$ and divide by n
(4) Square the grand total and divide by N
(5) The three sums of squares can now be formed quite easily, or alternatively S_2 can be found by subtracting S_1 from S

The analysis of variance table is now built up by entering the appropriate sums of squares and degrees of freedom as described earlier and dividing to form the mean squares.

1.8 Example

Suppose that it is desired to test the difference between four varieties of wheat, and that 10 plots have been sown with each variety, making 40 plots in all. The yields from these 40 plots would provide the data for a one-way classification with equal numbers of observations at each level.

In this case $p = 4$ and $n = 10$, and the resultant table would look like –

Source of estimate	Sums of squares	Degrees of freedom	Mean squares	Ratio
Between varieties	22·7	3	7·57	1·46
Within varieties	187·2	36	5·2	
Total	209·9	39		

Now this experiment must be examined under Model I since the four varieties are specially chosen by the experimenter as the ones in which he is particularly interested. Hence the factor under test will have a systematic effect. From tables we obtain the 5 per cent value of F with $(3, 36)$ degrees of freedom to be $2 \cdot 87$. As this value is greater than the calculated value of the ratio we conclude that there is no evidence of any real difference in the four varieties of wheat under test.

Alternatively, suppose we were to consider four machines, and that 10 articles were manufactured on each machine, making 40 articles in all. Some measurement on these articles would provide the data, and a similar table with the word "machines" replacing "varieties" might be the result. If the machines are regarded as a random sample from all possible machines, then this would be a case for Model II. But if they were to be tested by a machine user, then Model I would be appropriate, because the user would only be interested in the particular four machines to which he had access. The choice of the appropriate model can only be made by a person having some background knowledge regarding the purpose of the experiment.

CHAPTER 2
TWO-WAY CLASSIFICATIONS

2.1 Introduction

In investigations in which many factors are involved, it will not be economic or efficient to investigate the effect of one factor at a time on the particular result under investigation. Also, such a procedure gives no information about the possible interactions which may exist between relevant factors. Consequently, it is necessary to consider experiments in which more than one factor has been considered. In this chapter, we shall consider the case of two factors affecting the experimental results, and in the next chapter we shall generalise these results, first to the case of three factors and then to the more general case of many factors. When two factors affect the experiment, they may both have systematic effects (Model I) or random effects (Model II). If one has a systematic effect and the other is random we have an example of what we called the Mixed Model.

It will be convenient to distinguish between two types of two-way classification which have become known as "crossed" and "nested".

(i) *Crossed classification*

Let the two factors A and B have p and q levels respectively, and let there be n observations in each of the $p \times q$ cells of the two-way table, so that there will be $N = npq$ observations in all. The data are shown diagrammatically in Table 1. We may write the observations

$$x_{ij\alpha} = \mu + F_i + G_j + I_{ij} + \epsilon_{ij\alpha} \qquad (1)$$

$$(i = 1...p, \quad j = 1...q, \quad \alpha = 1...n)$$

where μ is the overall mean, F_i is the effect due to the ith level of the first factor, G_j is the effect due to the jth level of the second factor, I_{ij} is an interaction term representing the departure of the mean of the observations in the (ij)th cell from the sum of the first three terms of (1), and $\epsilon_{ij\alpha}$ takes into account the variation within a particular cell. As before, $\epsilon_{ij\alpha}$ will be assumed to be normally distributed about zero mean and with variance σ^2. We shall also assume that the expected values of $F_.$, $G_.$, $I_{i.}$, $I_{.j}$ are all zero. This is no restriction since, if they are not, then they can always be made so by adjustment of the

FACTOR A

FACTOR B	A_1	A_2	A_3	. . . . . .	A_p	Totals
B_1	$x_{111}, x_{112}, \ldots x_{11n}$	$x_{211}, x_{212}, \ldots x_{21n}$	$x_{311}, x_{312}, \ldots x_{31n}$		$x_{p11}, x_{p12}, \ldots x_{p1n}$	$p n\, x_{.1.}$
B_2	$x_{121}, x_{122}, \ldots x_{12n}$	$x_{221}, x_{222}, \ldots x_{22n}$	$x_{321}, x_{322}, \ldots x_{32n}$		$x_{p21}, x_{p22}, \ldots x_{p2n}$	$p n\, x_{.2.}$
B_3	$x_{131}, x_{132}, \ldots x_{13n}$	$x_{231}, x_{232}, \ldots x_{23n}$	$x_{331}, x_{332}, \ldots x_{33n}$		$x_{p31}, x_{p32}, \ldots x_{p3n}$	$p n\, x_{.3.}$
. . . . . .				. . . . . .		. . . . . .
B_q	$x_{1q1}, x_{1q2}, \ldots x_{1qn}$	$x_{2q1}, x_{2q2}, \ldots x_{2qn}$	$x_{3q1}, x_{3q2}, \ldots x_{3qn}$		$x_{pq1}, x_{pq2}, \ldots x_{pqn}$	$p n\, x_{.q.}$
Totals	$q n\, x_{1..}$	$q n\, x_{2..}$	$q n\, x_{3..}$	. . . . . .	$q n\, x_{p..}$	$p q n\, x_{...}$

Table 1 – Data for a two-way classification with n replications per cell

other factors.

It is also possible to assume that there is no interaction between the two factors, in which case a population of the form

$$x_{ij\alpha} = \mu + F_i + G_j + \epsilon_{ij\alpha} \qquad (2)$$

may be assumed. This form is usually employed in the case when there is only one observation per cell.

(ii) *Nested classification*

If one of the factors is such that no main effect can be attributed to it, then a different form for the observation must be assumed. An example of such a factor is where it corresponds to a sample number. Clearly, an assumption that all the samples having a particular number have a common effect is not possible. Such a factor is said to be "nested" within the main classification. Let factor B be nested within A, then for this type of data we assume that the observations take the form

$$x_{ij\alpha} = \mu + F_i + I_{j(i)} + \epsilon_{ij\alpha} \qquad (3)$$

Here the F_i term corresponds to the effect of the main factor, while the $I_{j(i)}$ corresponds to the effect of the nested factor for a particular value i of the main factor. As before, the $\epsilon_{ij\alpha}$ term corresponds to the variation within a particular cell and is assumed to have the same distribution.

We shall now consider these two types separately under our separate models.

2.2 Two crossed classifications with replication

We have already stated in (1) the form which we shall assume for the observations $x_{ij\alpha}$. Using the dot notation as before, we may write

$$x_{ij\alpha} - x_{...} = (x_{i..} - x_{...}) + (x_{.j.} - x_{...})$$
$$+ (x_{ij.} - x_{i..} - x_{.j.} + x_{...}) + (x_{ij\alpha} - x_{ij.})$$

Squaring both sides and summing over i, j and α we get

$$\sum_{ij\alpha}(x_{ij\alpha} - x_{...})^2 = nq\sum_i (x_{i..} - x_{...})^2 + np\sum_j (x_{.j.} - x_{...})^2$$
$$+ n\sum_{ij}(x_{ij.} - x_{i..} - x_{.j.} + x_{...})^2 + \sum_{ij\alpha}(x_{ij\alpha} - x_{ij.})^2 \qquad (4)$$

or

$$S = S_1 + S_2 + S_3 + S_4$$

The cross-product terms disappear as for the single factor case. Now S will have $N-1$ degrees of freedom, S_1 will have $p-1$ degrees of freedom, S_2 will have $q-1$ degrees of freedom, and S_4 will have $N-pq$ degrees of freedom since they are all computed from the deviations of observations from various sample means. In order to make the degrees of freedom equal on both sides of (4), S_3 must have $(p-1)(q-1)$ degrees of freedom. N is of course the total number of observations = npq.

Accordingly we get the following analysis of variance table:

Source of estimate	Sums of squares	Degrees of freedom	Mean squares
Factor A	$S_1 = nq\Sigma_i(x_{i..} - x_{...})^2$	$p-1$	$M_1 = S_1/(p-1)$
Factor B	$S_2 = np\Sigma_j(x_{.j.} - x_{...})^2$	$q-1$	$M_2 = S_2/(q-1)$
$A \times B$ interaction	$S_3 = n\Sigma_{ij}(x_{ij.} - x_{i..} - x_{.j.} + x_{...})^2$	$(p-1)(q-1)$	$M_3 = S_3/(p-1)(q-1)$
Within cells	$S_4 = \Sigma_{ija}(x_{ija} - x_{ij.})^2$	$N-pq$	$M_4 = S_4/(N-pq)$
Total	$S = \Sigma_{ija}(x_{ija} - x_{...})^2$	$N-1$	

Substituting from (1) into (4), the expected values we get are

$$S_1 = nq\sum_i (F_i + I_{i.} + \epsilon_{i..} - F_. - I_{..} - \epsilon_{...})^2$$

$$= nq\left\{\sum_i (F_i - F_.)^2 + \sum_i (I_{i.} - I_{..})^2 + \sum_i (\epsilon_{i..} - \epsilon_{...})^2\right\}$$

$$S_2 = np\left\{\sum_j (G_j - G_.)^2 + \sum_j (I_{.j} - I_{..})^2 + \sum_j (\epsilon_{.j.} - \epsilon_{...})^2\right\}$$

$$S_3 = n\sum_{ij}(I_{ij} - I_{i.} - I_{.j} + I_{..})^2 + n\sum_{ij}(\epsilon_{ij.} - \epsilon_{i..} - \epsilon_{.j.} + \epsilon_{...})^2$$

and

$$S_4 = \sum_{ija}(\epsilon_{ija} - \epsilon_{ij.})^2$$

The cross-product terms all vanish since the terms in (1) are assumed independent. Hence S_4 is an unbiased estimate of $pq(n-1)\sigma^2$, so that M_4 is an unbiased estimate of σ^2, the error variance within cells.

We must now investigate this experiment under our three specified models. Under Model I, the F's, G's and I's will be fixed quantities.

Hence

M_1 is an unbiased estimate of $\dfrac{nq}{(p-1)} \sum_i F_i^2 + \sigma^2$

M_2 " " " " " $\dfrac{np}{(q-1)} \sum_j G_j^2 + \sigma^2$

M_3 " " " " " $\dfrac{n}{(p-1)(q-1)} \sum_{ij} I_{ij}^2 + \sigma^2$

and M_4 " " " " " σ^2

The interaction terms disappear from the main effects since we have assumed that the expected values of $I_{i.}$ and $I_{.j}$ are zero. This means that, as we are now dealing with the whole populations of levels, $I_{i.} = I_{.j} = 0$. Now the existence of an interaction effect can be tested by comparing the ratio M_3/M_4 to the F-distribution with $(p-1)(q-1)$ and $(N-pq)$ degrees of freedom. Similarly the existence of main effects due to factors A and B can be tested by means of the ratios M_1/M_4 and M_2/M_4 respectively.

If a non-significant value of M_3/M_4 occurs, some people prefer to sum the last two rows of the analysis of variance table and get a better estimate of σ^2 as

$$(S_3 + S_4) \Big/ \{(N-pq) + (p-1)(q-1)\}$$

This is not to be recommended since all the F-test can do is to cause the rejection of the hypothesis that the interaction term is zero if a significant value is obtained. A non-significant value does not provide evidence that the hypothesis is true. In other words, there is always the possibility of some interaction being present and not showing up on the F-test. Hence, the best estimate of σ^2 is always $S_4/(N-pq)$, unless some external evidence about the experiment is available.

If we now turn to examine this case under Model II, the F's, G's and I's are assumed to be a random sample from some specified populations with zero means and variances σ_A^2, σ_B^2 and σ_I^2 respectively. Again these populations are assumed to be normally distributed. Hence we may write

M_1 is an unbiased estimate of $nq\sigma_A^2 + n\sigma_I^2 + \sigma^2$

M_2 " " " " " $np\sigma_B^2 + n\sigma_I^2 + \sigma^2$

M_3 is an unbiased estimate of $n\sigma_I^2 + \sigma^2$

M_4 " "   "   "   " σ^2

As before, we may test the hypothesis that there is no interaction term by comparing the ratio M_3/M_4 to the F-distribution with the same degrees of freedom. However, in order to test for the main effects of the two factors, their respective mean square must now be divided by the interaction mean square and not by the "within cells" or residual mean square, as was the case under Model I. This is one of the most important differences between the two models. When a factor has a random effect, then the main effect is tested by the interaction sum of squares, whereas if it has a systematic effect then one must divide by the residual sum of squares. This rule, however, is not general, as will be seen when we consider the Mixed Model.

In this case σ^2 is estimated by M_4, σ_I^2 by $\frac{1}{n}(M_3 - M_4)$, σ_B^2 by $\frac{1}{np}(M_2 - M_3)$ and σ_A^2 by $\frac{1}{nq}(M_1 - M_3)$.

Under the Mixed Model we shall consider factor A to have a random effect and factor B to have the systematic effect. As under Model II, the F's and I's are assumed to be a random sample from some specified populations which are normally distributed with zero means and variances σ_A^2 and σ_I^2. As before, $G_.$, the mean of the effects of factor B, can be taken to be zero. Hence we find

M_1 is an unbiased estimate of $nq\sigma_A^2 + \sigma^2$

M_2 " "   "   "   "   " $np\sum_j G_j^2/(q-1) + n\sigma_I^2 + \sigma^2$

M_3 " "   "   "   "   " $n\sigma_I^2 + \sigma^2$

M_4 " "   "   "   "   " σ^2

We note that the interaction term disappears from M_1 because for a fixed value of i, all the values of I_{ij} in the population have been taken into the sample. Hence we may assume that $I_{i.} = 0$. This means that we must define the I's as being normally distributed only for fixed values of the j suffix.

Again the interaction term is tested by comparing the ratio M_3/M_4 to the F-distribution with the same degrees of freedom. The existence of the effects of random factor A is tested by dividing the mean square M_1 by the "within cells" or residual mean square. The test for the main effect of factor B is obtained by dividing M_2 by the interaction mean square. Thus the tests for the main effects in the mixed model work conversely to the tests of Models I and II. Here the factor with the random effect is tested by dividing by the residual mean square, whereas if this factor had occurred under Model II, then the test would be made by dividing by the interaction sum of squares. A similar inversion occurs when the factor with fixed effects is considered under Model I and the Mixed Model.

Now σ^2 is estimated by M_4, σ_I^2 by $\frac{1}{n}(M_3 - M_4)$ and σ_A^2 by $\frac{1}{nq}(M_1 - M_4)$.

2.3 Two crossed classifications with no replication

A special case of the previous paragraph occurs when there is only one observation made in each of the cells, i.e. $n = 1$. With $n = 1$, the form for the observations becomes

$$x_{ij} = \mu + F_i + G_j + I_{ij} + \epsilon_{ij} \qquad (5)$$

and the following analysis of variance table is derived:

Source of estimate	Sums of squares	Degrees of freedom	Mean squares
Factor A	$S_1 = q \sum_i (x_{i.} - x_{..})^2$	$p - 1$	$M_1 = S_1/(p-1)$
Factor B	$S_2 = p \sum_j (x_{.j} - x_{..})^2$	$q - 1$	$M_2 = S_2/(q-1)$
Interaction	$S_3 = \sum_{ij} (x_{ij} - x_{i.} - x_{.j} + x_{..})^2$	$(p-1)(q-1)$	$M_3 = S_3/(p-1)(q-1)$
Total	$S = \sum_{ij} (x_{ij} - x_{..})^2$	$N - 1$	

No "Within Cells" sum of squares is available in this case as there is only one observation in each cell.

Under Model I,

M_1 is an unbiased estimate of $q \sum_i F_i^2/(p-1) + \sigma^2$

M_2 " " " " " $p \sum_j G_j^2/(q-1) + \sigma^2$

M_3 " " " " " $\sum_{ij} I_{ij}^2/(p-1)(q-1) + \sigma^2$

and hence no tests are possible, for the assumption that either of the main effects or the interaction effect is zero gives us no suitable mean squares to compare.

However, under Model II,

M_1 is an unbiased estimate of $q\sigma_A^2 + \sigma_I^2 + \sigma^2$

M_2 " " " " " $p\sigma_B^2 + \sigma_I^2 + \sigma^2$

M_3 " " " " " $\sigma_I^2 + \sigma^2$

and it is possible to test for the main effects by dividing their respective mean squares by the interaction mean square.

Finally, it may be seen that, under the Mixed Model of section 2.2 only the existence of the fixed effects of factor B can be tested by dividing M_2 by the interaction mean square.

Thus it is seen that to assume a population of the form of (5) does not lead to a simple set of tests. Indeed, only in certain cases are any tests available at all. Thus the most frequent form of the population, used in the case when there is only a single observation in each cell, is the one where the interaction is assumed to be zero. This will be described in the next section.

2.4 Two crossed classifications with no interaction

When it is assumed that there is no interaction between the two factors, the population takes the form of (2). This is of importance when there is only a single observation in each cell. If it is applied when n is not equal to 1, the two mean squares corresponding to "interaction" and "within cells" merely provide independent estimates of σ^2, and hence they may be combined to give a more reliable estimate of σ^2 with $(N - pq) + (p - 1)(q - 1)$ degrees of freedom. In fact, the analyst has deliberately avoided the test for interaction effects by adding the "$A \times B$ interaction" and "within cells" lines in the analysis of variance table. Tests for existence of the main effects are obtained by dividing the appropriate mean square by the mean square obtained from this new "within cells" line in the table. These tests are the same no matter which model is appropriate to the particular experiment.

When there is no replication (2) takes the form

$$x_{ij} = \mu + F_i + G_j + \epsilon_{ij} \tag{6}$$

and the same analysis of variance table ensues as in section 2.3, except that the source of S_3 must no longer be termed "interaction" but, more rightly, may be called "error", so that S_3 is spoken of as the error sum of squares.

Instead of referring to our three models it is only necessary in this case to say that

M_1 is an unbiased estimate of $\theta + \sigma^2$
M_2 " "   "   "   " $\phi + \sigma^2$
M_3 " "   "   "   " σ^2

where $\theta = q \Sigma_i F_i^2 /(p - 1)$ if A is a factor having fixed effects
       $= q \sigma_A^2$ "   "   "   "   "   random   "

and $\phi = p\Sigma_j G_j^2/(q-1)$ if B is a factor having fixed effects
$\qquad = p\sigma_B^2$ " " " "   "   " random   "

Irrespective of the nature of the two factors involved in this case, and thus of the type of model assumed, the tests for the existence of the main effects are obtained by dividing the appropriate mean square by M_3 and comparing the ratio with the F-distribution. Thus a population of the form of (6) is much more useful in the case of single observations in the cells than was the assumption (5) in section 2.3.

2.5 Two nested classifications

In section 2.1 we have described an alternative system of two-way classification where one of the factors is "nested" within the other factor. In this case, no main effect can be associated with the nested factor, and the appropriate form to be assumed in this case was given in equation (3), assuming factor B to be nested within A.

We shall now take as our basic equation

$$x_{ij\alpha} - x_{...} = (x_{i..} - x_{...}) + (x_{ij.} - x_{i..}) + (x_{ij\alpha} - x_{ij.})$$

As factor B is assumed not to have a main effect, no term of the form $(x_{.j.} - x_{...})$ need be included in the above equation. Squaring both sides and summing over i, j and α, we get

$$\sum_{ij\alpha}(x_{ij\alpha} - x_{...})^2 = nq\sum_i(x_{i..} - x_{...})^2 + n\sum_{ij}(x_{ij.} - x_{i..})^2$$
$$+ \sum_{ij\alpha}(x_{ij\alpha} - x_{ij.})^2 \qquad (7)$$

or $\qquad S = S_1 + S_2 + S_3$

The cross-product terms again disappear on summation. Now S has $N-1$ degrees of freedom, S_1 has $p-1$ degrees of freedom and S_3 has $N-pq$ degrees of freedom since they are all computed from the deviations of observations from various sample means. To balance the degrees of freedom on both sides of (7), S_2 must have $p(q-1)$ degrees.

Hence the following analysis of variance table can be drawn up:

Source of estimate	Sums of squares	Degrees of freedom	Mean squares
Factor A	$S_1 = nq\sum_i(x_{i..} - x_{...})^2$	$(p-1)$	$M_1 = S_1/(p-1)$
Between cells (within factor A)	$S_2 = n\sum_{ij}(x_{ij.} - x_{i..})^2$	$p(q-1)$	$M_2 = S_2/p(q-1)$
Within cells	$S_3 = \sum_{ij\alpha}(x_{ij\alpha} - x_{ij.})^2$	$N-pq$	$M_3 = S_3/(N-pq)$
Total	$S = \sum_{ij\alpha}(x_{ij\alpha} - x_{...})^2$	$N-1$	

The second line of the table now represents the variation between the cells of the two-way table which correspond to fixed levels of factor A. As before, the next line represents the variation within the cells due to repeated trials of the experiment for fixed levels of the two factors A and B.

Substituting from (3) into (7), the expected values we get are

$$S_1 = nq \sum_i (F_i + I_{.(i)} + \epsilon_{i..} - F_. - I_{.(.)} - \epsilon_{...})^2$$
$$= nq \left\{ \sum_i (F_i - F_.)^2 + \sum_i (I_{.(i)} - I_{.(.)})^2 + \sum_i (\epsilon_{i..} - \epsilon_{...})^2 \right\}$$
$$S_2 = n \left\{ \sum_{ij} (I_{j(i)} - I_{.(i)})^2 + \sum_{ij} (\epsilon_{ij.} - \epsilon_{i..})^2 \right\}$$
$$S_3 = \sum_{ija} (\epsilon_{ija} - \epsilon_{ij.})^2$$

The cross-product terms all vanish since the terms in (3) are assumed independent. Hence S_3 is again an unbiased estimate of $pq(n-1)\sigma^2$, so that M_3 is an unbiased estimate of σ^2, the error variance within cells.

Under Model I, the F's and I's will be fixed quantities and hence M_1, M_2 and M_3 become unbiased estimates of

$$\frac{nq}{(p-1)} \sum_i F_i^2 + \sigma^2$$

$$\frac{n}{p(q-1)} \sum_{ij} I_{j(i)}^2 + \sigma^2$$

and σ^2 respectively.

Thus, the existence of the main effect of factor A and the nested effect of factor B are both tested by dividing the appropriate mean square by M_3 and referring to the F-distribution.

If we assume factors A and B to have random effects, then Model II is appropriate and M_1, M_2 and M_3 become unbiased estimates of

$$nq\sigma_A^2 + n\sigma_I^2 + \sigma^2$$

$$n\sigma_I^2 + \sigma^2$$

and σ^2 respectively.

Hence the existence of the main effect of factor A is now tested by means of the ratio M_1/M_2 and the existence of σ_I^2 is tested for by dividing M_2 by M_3. Now σ^2 is estimated by M_3, σ_I^2 by $\frac{1}{n}(M_2 - M_3)$ and σ_A^2 by $\frac{1}{nq}(M_1 - M_2)$.

The analysis of the Mixed Model with factor A having systematic effects and factor B having random effects is similar to Model II if we replace σ_F^2 by $\Sigma_i F_i^2/(p-1)$. If the effects of the two factors are reversed, then the analysis corresponds to Model I with $\Sigma_i F_i^2/(p-1)$ replaced by σ_F^2.

2.6 Missing data

In all the types of two-way classification which have been considered, it has been assumed that there has been the same number of replications of the experiment in each cell of the table. If this is not true, then it is usually impossible to separate the overall sums of squares into independent components due to main effects and interaction terms. A method of testing such data for main effects and interaction is available, but as it presents certain difficulties both in theory and computation it has not been included in this book.

Care should always be taken to ensure that the number of observations in each cell is identical, but, even with the greatest care, it will happen occasionally that a few observations are lost. If this occurs, it is recommended that the following approximate procedure be adopted in preference to the difficult analysis which has been omitted. Although approximate, the significance tests obtained by this method will not be seriously in error unless something like 20 per cent of the data is missing. The method consists of inserting estimates of the missing observations which have been chosen to minimise the residual or "within cells" variance and then completing the analysis as for a complete set of data. This is equivalent to replacing the missing observation by the mean of the other observations in the cell. Care should be taken, however, not to include these estimates when computing the relevant degrees of freedom.

If it is a design with no replications and some of the data are missing, i.e. some of the cells are completely empty, then the above approximate method will not work and a more sophisticated method is needed (see for example Bennett and Franklin, p. 382). It should be stressed that the substitution of estimates for missing data does not in any way recover the information that is lost through lack of data, but is merely a computational device to enable the easy computations that apply to complete data to be used when the data are incomplete.

All these difficulties and approximations can be avoided if it is remembered that analysis of variance does not take kindly to incomplete sets of data, and if care is taken to make the experimental data complete. The situation is summed up by Cochran and Cox (1957) when they say "the only solution to the missing data problem is not to have them."

2.7 Computational note

As for the single-factor case, the various sums of squares used in this chapter can be expressed in alternative forms which are more suitable for computation. By completing the squares we obtain for the crossed classification with replication

$$S = \sum_{ija}(x_{ija} - x_{...})^2 = \left| \sum_{ija} x_{ija}^2 - \left(\sum_{ija} x_{ija} \right)^2 \middle/ N \right.$$

$$S_1 = nq \sum_i (x_{i..} - x_{...})^2 = \sum_i T_{i.}^2/nq - \left(\sum_{ija} x_{ija} \right)^2 \middle/ N$$

$$S_2 = np \sum_j (x_{.j.} - x_{...})^2 = \sum_j T_{.j}^2/np - \left(\sum_{ija} x_{ija} \right)^2 \middle/ N$$

$$S_3 = n \sum_{ij} (x_{ij.} - x_{i..} - x_{.j.} + x_{...})^2$$
$$= \sum_{ij} T_{ij}^2/n - \sum_i T_{i.}^2/nq - \sum_j T_{.j}^2/np + \left(\sum_{ija} x_{ija} \right)^2 \middle/ N$$

$$S_4 = \sum_{ija}(x_{ija} - x_{ij.})^2 = \sum_{ija} x_{ija}^2 - \sum_{ij} T_{ij}^2/n$$

where T_{ij} is the total of the observations in the (ij)th cell and $T_{i.}$ is the total of the observations in the ith classification, and similarly $T_{.j}$ is the total of the observations in the jth classification. The appropriate sums of squares can now be built up by forming the individual terms on the right-hand side of the above equations. S_3 is usually calculated by subtracting $(S_1 + S_2 + S_4)$ from S.

For the nested classifications we replace S_2 above by

$$S_2 = n \sum_{ij} (x_{ij.} - x_{i..})^2 = \sum_{ij} T_{ij}^2/n - \sum_i T_{i.}^2/nq$$

and S_3 is not required.

Thus in both cases it is necessary to form the sums within each cell, T_{ij}, and the sums over rows and columns, $T_{i.}$ and $T_{.j}$ of our two-way table. A check should be made to see that $\Sigma_i T_{i.} = \Sigma_j T_{.j}$ which is, of course, the grand total $\Sigma_{ija} x_{ija}$. When these subsidiary totals have all been found then it is a matter of routine to find the appropriate sums of squares for use in the above formula. A desk calculating machine is recommended if the quantity of data is large, Such an instrument has the advantage that totals and sums of squares can be formed at the same time, which provides some useful checks on the previous work. If a desk machine is not available then at least some of the labour may be eased by the use of the tables of squares which are readily obtainable.

2.8 Example 1 – Crossed classification

Consider the following example from Crump (1946). The data are drawn from a series of genetical experiments on egg production and comprise the total number of eggs laid by each of 12 females from 25 races of *Drosophila melanogaster*, on the fourth day of laying, the whole experiment being carried out 4 times.

Here our two factors are Race and Experiment, and the experimental units which provide the replication are the females. Hence we may put $p = 25$, $q = 4$, and $n = 12$, and the following table is obtained:

Source of variation	Degrees of freedom	Mean squares	Quantities of which mean squares are unbiased estimates	Ratios
Races (A)	24	$M_1 = 46{,}659$	$48\sigma_A^2 + 12\sigma_I^2 + \sigma^2$	$M_1/M_3 = 102$
Experiments (B)	3	$M_2 = 3{,}243$	$300\sigma_B^2 + 12\sigma_I^2 + \sigma^2$	$M_2/M_3 = 7\cdot 1$
$A \times B$ interaction	72	$M_3 = 459$	$12\sigma_I^2 + \sigma^2$	$M_3/M_4 = 1\cdot 99$
Within cells	1100	$M_4 = 231$	σ^2	
Total	1199			

This is an example of two crossed classifications with replication, and as both factors may be regarded as having effects which are randomly selected from a large population of such effects, the data must be analysed under Model II.

Accordingly the appropriate average values of the mean squares have been entered in column 4 of the table. This enables the analyst to decide which are the appropriate ratios that are required.

The existence of the interaction is tested by comparing the ratio $1\cdot 99$ with the F-distribution with $(72, 1100)$ degrees of freedom which is significant at the 1 per cent level of significance. Hence there is strong evidence that there is interaction present. The existence of a main effect due to experiments is tested by comparing the ratio $7\cdot 1$ with the F-distribution with $(3, 72)$ degrees of freedom which is again significant at the 1 per cent level. Similarly the other main effect is tested by comparing the ratio 102 with the F-distribution with $(24, 72)$ degrees of freedom, and this again is significant. Thus we conclude that there is a marked difference between the experiments and between the races and also some interaction between the two factors. Estimates of the variances involved give $\sigma^2 = 231$, $\sigma_I^2 = 19$, $\sigma_B^2 = 9\cdot 3$ and $\sigma_A^2 = 962$.

If instead of regarding the 25 races used as a random selection of a large number of possible races, we were to regard the 25 races as being specially chosen, i.e. they were the 25 in which the experimenter was particularly interested, then the appropriate analysis would be a Mixed Model with factor A (races) having a fixed effect and factor B (experiments) having a systematic effect. The last two columns of the table would then become

Quantities of which mean squares are unbiased estimates	Ratios
$2\sum_i F_i^2 + 12\sigma_I^2 + \sigma^2$	$M_1/M_3 = 102$
$300\sigma_B^2 + \sigma^2$	$M_2/M_4 = 14$
$12\sigma_I^2 + \sigma^2$	$M_3/M_4 = 1\cdot 99$
σ^2	

Thus, in order to test for the main effect of experiments we must now use the ratio M_2/M_4 and not M_2/M_3 as previously. Also this new ratio must be compared with the F-distribution with (3, 1100) degrees of freedom, where it will again be significant. All the other tests follow as before. The estimate of σ_B^2 now becomes 10.

In this example, we have arrived at the same conclusions whether we assumed the races to have a random or a systematic effect. This will not be true of every example, and care must be exercised in choosing the correct method of analysis. This is done by first deciding if the factors are nested or crossed and then deciding if they are to be regarded as random or systematic. In the present case, the experiments themselves must be regarded as a random sample from all the experiments that it is possible to carry out, and hence we cannot assume them to have a systematic effect.

2.9 Example 2 – Nested classification

The data for a nested classification are similar to those for a single factor with replications except that the replications fall into groups and it is necessary to find out if there is any variation between the groups for a given level of the main factor. A suitable example of this type of data is given by Wernimont (1947). The experiment was designed to test the homogeneity of the copper content of a series of bronze castings from the same pour. Two samples were taken from each of eleven castings and each sample was analysed in duplicate.

The two samples are chosen at random from each casting, and hence there can be no relationship between the first of the two samples analysed for one casting and the first sample examined from the second casting. Hence, sample is a nested factor inside the main factor which is castings. The analysis of section 2.5 yields the following analysis of variance table.

Source of variation	Sums of squares	Degrees of freedom	Mean squares	Quantities of which mean squares are unbiased estimates
Between castings	3·2031	10	0·3203	$4\sigma_A^2 + 2\sigma_I^2 + \sigma^2$
Between samples (within castings)	1·9003	11	0·1728	$2\sigma_I^2 + \sigma^2$
Within samples	0·0351	22	0·0016	σ^2
Total	5·1385	43		

The last column has been completed as for a Model II analysis because both the samples from the castings and the castings themselves may be regarded as a random sample from some population of samples or castings.

The existence of variations in the copper content between samples from a given casting is tested by comparing the ratio $\frac{0·1728}{0·0016} = 108·0$ with the F-distribution with (11, 22) degrees of freedom, the result being highly significant. The existence of variation in the copper content between the different castings is tested by comparing the ratio $\frac{0·3203}{0·1728} = 1·853$ with the F-distribution with (10, 11) degrees of freedom, the result being not significant at the 5 per cent level. We would thus conclude that there is a marked difference between samples from a given casting but not between castings.

CHAPTER 3

GREATER NUMBERS OF CLASSIFICATIONS

3.1 Introduction

The methods of the previous chapter may be readily extended to experiments in which more than two factors are involved. These factors may be either crossed or nested as before. A third possibility now arises in that some experiments may include both crossed and nested classifications. We shall first consider the case where all the factors are crossed with replications, which is an extension of section 2.2, and then examine the case in which each factor is nested within the previous classification, which is an extension of section 2.5. Finally we shall consider the complex case where both types of factors are present. Each case will be examined under our three specified models.

3.2 Crossed classifications with replications

It will be sufficient indication of the general case to consider three factors only which we shall call A, B and C. The appropriate model to represent the data $x_{ijk\alpha}$, when the factors have p, q and r levels and there are n replicates in each cell, may be written

$$x_{ijk\alpha} = \mu + F_i + G_j + H_k + I_{ij} + J_{ik} + K_{jk} + L_{ijk} + \epsilon_{ijk\alpha} \qquad (1)$$
$$(i = 1...p,\ j = 1...q,\ k = 1...r,\ \alpha = 1...n\ \text{and}\ N = npqr)$$

where F_i, G_j and H_k stand for the main effects of the factors, I_{ij}, J_{ik} and K_{jk} stand for any possible interactions between pairs of factors, and L_{ijk} represents the possible interaction between all three factors. It may readily be seen that the number of main effects and interactions to be tested for is equal to $(2^m - 1)$ where m is the number of factors in the experiment.

As before, we can produce a partitioning of the overall sum of squares to form the analysis of variance table shown opposite, and the mean squares M_1 to M_8 are obtained by dividing the sums of squares S_1 to S_8 by their respective degrees of freedom. Again $\epsilon_{ijk\alpha}$ is assumed to be normally distributed about zero mean and with variance σ^2. Also we shall assume that if any of the terms F_i, G_j...L_{ijk} are averaged over one or more suffices then the expected value of that term

Source of estimate	Sums of squares	Degrees of freedom
Main effects		
Factor A	$S_1 = nqr \sum_i (x_{i...} - x_{....})^2$	$p - 1$
Factor B	$S_2 = npr \sum_j (x_{.j..} - x_{....})^2$	$q - 1$
Factor C	$S_3 = npq \sum_k (x_{..k.} - x_{....})^2$	$r - 1$
Two-factor interaction		
$A \times B$	$S_4 = nr \sum_{ij} (x_{ij..} - x_{i...} - x_{.j..} + x_{....})^2$	$(p-1)(q-1)$
$A \times C$	$S_5 = nq \sum_{ik} (x_{i.k.} - x_{i...} - x_{..k.} + x_{....})^2$	$(p-1)(r-1)$
$B \times C$	$S_6 = np \sum_{jk} (x_{.jk.} - x_{.j..} - x_{..k.} + x_{....})^2$	$(q-1)(r-1)$
Three-factor interaction		
$A \times B \times C$	$S_7 = n \sum_{ijk} (x_{ijk.} - x_{ij..} - x_{i.k.} - x_{.jk.} + x_{i...} + x_{.j..} + x_{..k.} - x_{....})^2$	$(p-1)(q-1)(r-1)$
Within cells (error)	$S_8 = \sum_{ijk\alpha} (x_{ijk\alpha} - x_{ijk.})^2$	$N - pqr$
Total	$S = \sum_{ijk\alpha} (x_{ijk\alpha} - x_{....})^2$	$N - 1$

is zero. This can always be ensured by suitable adjustment of the other factors. Bearing these assumptions in mind, it is possible to show that, under Model 1, M_1 to M_8 will be unbiased estimates of

$$nqr \Sigma_i F_i^2/(p-1) + \sigma^2$$
$$npr \Sigma_j G_j^2/(q-1) + \sigma^2$$
$$npq \Sigma_k H_k^2/(r-1) + \sigma^2$$
$$nr \Sigma_{ij} I_{ij}^2/(p-1)(q-1) + \sigma^2$$
$$nq \Sigma_{ik} J_{ik}^2/(p-1)(r-1) + \sigma^2$$
$$np \Sigma_{jk} K_{jk}^2/(q-1)(r-1) + \sigma^2$$
$$n \Sigma_{ijk} L_{ijk}^2/(p-1)(q-1)(r-1) + \sigma^2$$

and σ^2 respectively.

Hence, in order to test the existence of any main effect or interaction, the appropriate mean square should be divided by the error mean square and the result referred to the F-distribution with the appropriate degrees of freedom.

To examine this case under Model 2 we must first assume that the F's, G's ... and L's are random samples from populations which are normally distributed with zero means and variances σ_F^2, σ_G^2, ... σ_L^2 respectively. Hence M_1 to M_8 will be unbiased estimates of

$$nqr\sigma_F^2 + nr\sigma_I^2 + nq\sigma_J^2 + n\sigma_L^2 + \sigma^2$$
$$npr\sigma_G^2 + np\sigma_K^2 + nr\sigma_I^2 + n\sigma_L^2 + \sigma^2$$
$$npq\sigma_H^2 + np\sigma_K^2 + nq\sigma_J^2 + n\sigma_L^2 + \sigma^2$$
$$nr\sigma_I^2 + n\sigma_L^2 + \sigma^2$$
$$nq\sigma_J^2 + n\sigma_L^2 + \sigma^2$$
$$np\sigma_K^2 + n\sigma_L^2 + \sigma^2$$
$$n\sigma_L^2 + \sigma^2$$

and σ^2 respectively.

The tests for the existence of significant effects should begin with the highest-order interaction. The test for the three-factor interaction is made by dividing M_7 by the error mean square and referring the result to the F-distribution; and the test for the two-factor interaction is made by dividing the two-factor mean squares by M_7. However, unless the experimenter is prepared to assume, either on the basis of some external evidence or by a non-significant result in the previous test, that one or more of the two-factor interactions are zero, no accurate tests for main effects are available. For example, to test the existence of the main effect of factor A, M_1 must be compared with a quantity which has an unbiased estimate equal to $(nr\sigma_I^2 + nq\sigma_J^2 + n\sigma_L^2 + \sigma^2)$ and this does not occur naturally in the analysis of variance table. Such a quantity can however be built up by taking $(M_4 + M_5 - M_7)$, but this does not have a mean square distribution. It may be taken to approximate to a mean square distribution with degrees of freedom equal to

$$g = \frac{(M_4 + M_5 - M_7)^2}{\dfrac{M_4^2}{f_4} + \dfrac{M_5^2}{f_5} + \dfrac{M_7^2}{f_7}}$$

where f_4, f_5 and f_7 are degrees of freedom defined on the analysis of variance table corresponding to S_4, S_5 and S_7. Thus if both the $A \times B$ and $A \times C$ interactions are significant then an approximate test for the main effect of factor A is to refer the ratio $M_1/(M_4+M_5-M_7)$ to the F-distribution with f_1 and g degrees of freedom. This method of approximating to a linear combination of mean squares by means of another mean square with some new degree of freedom is due to Satterthwaite (1946) and Welch (1936). If either the $A \times B$ or $A \times C$ interaction can be assumed to be zero, accurate tests for the main effect of factor A are obtained by referring the ratio M_1/M_5 or the ratio M_1/M_4 to the F-distribution with the appropriate degrees of freedom given in the table.

Analysis of this particular experiment under the Mixed Model should now be a simple task for the reader. Let us suppose that factors A and B have systematic effects and C has a random effect. Then M_1 to M_8 will be unbiased estimates of

$$nqr \sum_i F_i^2/(p-1) + nq\sigma_J^2 + \sigma^2$$

$$npr \sum_j G_j^2/(q-1) + np\sigma_K^2 + \sigma^2$$

$$npq\, \sigma_H^2 + \sigma^2$$

$$nr \sum_{ij} I_{ij}^2/(p-1)(q-1) + n\sigma_L^2 + \sigma^2$$

$$nq\sigma_J^2 + \sigma^2$$

$$np\sigma_K^2 + \sigma^2$$

$$n\sigma_L^2 + \sigma^2$$

$$\text{and} \quad \sigma^2 \quad \text{respectively.}$$

Various terms have disappeared from these expected values of M_1 to M_8 because all the values in the populations for factors A and B have been used in this experiment. Hence we may assume that any of the terms F_i, G_j, ... , L_{ijk} which occur with a dot in place of the i or j suffix will be equal to zero. In this case, accurate tests are available for all main effects and interaction effects. To work out the appropriate test: first consider the row of the table corresponding to the effect which it is desired to test. The appropriate value of M has an expected value which is listed above. Another expected value must then be taken from the above list which is equal to the first except that the term corresponding to the effect being tested is omitted. The ratio of the two values of M is then referred to the F-table as before. For example, M_1/M_6 will test for the main effect of factor A; M_3/M_8 will test for the main effect of factor C; and M_4/M_7 will test for the $A \times B$ interaction.

It is left as an exercise to the reader to work out the unbiased estimates of M_1 to M_8 when only factor A has a systematic effect and B and C have random effects. The other cases, falling under the Mixed Model, can be dealt with by analogy.

3.3 Nested classifications

Examples of completely nested classifications occur frequently in the chemical industry where raw material is first broken up into batches and then into sub-batches.

Let us consider three factors where B is nested within A, and C is nested within B. As in section 2.5, no main effect can be associated with either of the nested factors, and thus, for this type of data we assume the observations take the form

$$x_{ijk\alpha} = \mu + F_i + I_{j(i)} + J_{k(ij)} + \epsilon_{ijk\alpha} \tag{2}$$

Here, the F_i term corresponds to the effect of factor A which is the main factor, the $I_{j(i)}$ correspond to the effect of the nested factor B for a particular value i of the main factor A, and the $J_{k(ij)}$ correspond to the effect of the nested factor C for particular values of factors A and B. The $\epsilon_{ijk\alpha}$ represents the variation within each cell and is assumed to be normally distributed about zero mean and with variance σ^2. The overall sum of squares can be partitioned to provide the analysis of variance table given on the opposite page. The second line of the table represents the variation in factor B for a given level of factor A; and the next line represents the variation in factor C for given levels of factors A and B.

When all the factors have systematic effects, Model 1 is applicable to these data. Substituting from (2) into S_1 to S_4 we obtain

$$nqr \Sigma_i F_i^2/(p-1) + \sigma^2$$
$$nr \Sigma_{ij} I_{j(i)}^2/p(q-1) + \sigma^2$$
$$n \Sigma_{ijk} J_{k(ij)}^2/pq(r-1) + \sigma^2$$
$$\text{and } \sigma^2$$

as unbiased estimates of M_1 to M_4. Tests for the effects of all three factors can be made by dividing the appropriate mean square by M_4. Under Model II with random factors we get

$$qrn\,\sigma_F^2 + rn\,\sigma_I^2 + n\,\sigma_J^2 + \sigma^2$$
$$rn\,\sigma_I^2 + n\,\sigma_J^2 + \sigma^2$$
$$n\,\sigma_J^2 + \sigma^2$$
$$\text{and } \sigma^2$$

as unbiased estimates of M_1 to M_4. We have here assumed that F_i, $I_{j(i)}$ and $J_{k(ij)}$ are normally distributed with zero mean and variances σ_F^2, σ_I^2 and σ_J^2 respectively. Tests for the existence of the main effect of factor A and the two nested factors all exist; factor A is tested by means of the ratio M_1/M_2, factor B by the ratio M_2/M_3, and factor C by the ratio M_3/M_4.

Source of estimate	Sums of squares	Degrees of freedom	Mean squares
Factor A	$S_1 = nqr \Sigma_i (x_{i...} - x_{....})^2$	$(p-1)$	$M_1 = \dfrac{S_1}{(p-1)}$
Factor B (within A)	$S_2 = nr \Sigma_{ij} (x_{ij..} - x_{i...})^2$	$p(q-1)$	$M_2 = \dfrac{S_2}{p(q-1)}$
Factor C (within A and B)	$S_3 = n \Sigma_{ijk} (x_{ijk.} - x_{ij..})^2$	$pq(r-1)$	$M_3 = \dfrac{S_3}{pq(r-1)}$
Within cells	$S_4 = \Sigma_{ijka} (x_{ijka} - x_{ijk.})^2$	$pqr(n-1)$	$M_4 = \dfrac{S_4}{pqr(n-1)}$
Total	$S = \Sigma_{ijka} (x_{ijka} - x_{....})^2$	$N-1$	

If we assume a Mixed Model with factor A having a systematic effect and factors B and C having random effects, then the analysis is the same as under Model 2 except that σ_F^2 is replaced by $\Sigma_i F_i^2/(p-1)$. If factor A is the only one having a random effect, then the analysis is the same as under Model 1 except that $\Sigma_i F_i^2/(p-1)$ is replaced by σ_F^2. The tests for existence of variation due to the three factors follow just as before. The remaining four cases which fall under the Mixed Model, i.e. those in which factors B and C have opposite effects, can be left as an exercise when it is pointed out that the term involving $I_{j(i)}$ disappears from the estimate of M_1 when factor B has a systematic effect, and the term involving $J_{k(ij)}$ disappears from the estimates of M_1 and M_2 when factor C has a systematic effect. Thus special care is needed when deciding on the appropriate tests to make.

3.4 Analyses involving nested and crossed classifications

When experiments contain both types of classification, the method of analysis depends on the order in which the criteria of classification occur.

Let us first consider the case where factors A and B are crossed and factor C is nested within factors A and B. Then the appropriate model for this type of data would be

$$x_{ijk\alpha} = \mu + F_i + G_j + I_{ij} + J_{k(ij)} + \epsilon_{ijk\alpha} \qquad (3)$$

where F_i and G_j represent the main effects of factors A and B (C will not have a main effect as it is nested), I_{ij} represents the interaction between A and B, and $J_{k(ij)}$ represents the effect of factor C for given values of the other two factors. The usual assumptions are made regarding the distribution of the error term $\epsilon_{ijk\alpha}$. The overall sum of squares is partitioned to provide the following analysis of variance table. If this is compared with the table of the previous section, it will

Source of estimate	Sums of squares	Degrees of freedom	Mean squares
Factor A	$S_1 = nqr \sum_i (x_{i...} - x_{....})^2$	$(p-1)$	$M_1 = \dfrac{S_1}{(p-1)}$
Factor B	$S_2 = npr \sum_j (x_{.j..} - x_{....})^2$	$(q-1)$	$M_2 = \dfrac{S_2}{(q-1)}$
$A \times B$ interaction	$S_3 = nr \sum_{ij} (x_{ij..} - x_{i...} - x_{.j..} + x_{....})^2$	$(p-1)(q-1)$	$M_3 = \dfrac{S_3}{(p-1)(q-1)}$
Factor C (within A and B)	$S_4 = n \sum_{ijk} (x_{ijk.} - x_{ij..})^2$	$pq(r-1)$	$M_4 = \dfrac{S_4}{pq(r-1)}$
Within cells	$S_5 = \sum_{ijk\alpha} (x_{ijk\alpha} - x_{ijk.})^2$	$pqr(n-1)$	$M_5 = \dfrac{S_5}{pqr(n-1)}$
Total	$S = \sum_{ijk\alpha} (x_{ijk\alpha} - x_{....})^2$	$N-1$	

be seen that the line corresponding to nested factor B has split into two lines to provide the main effect for the crossed factor B and an $A \times B$ interaction term. We must first consider this case under Model I with all the factors having systematic effects. Substituting from (3) into S_1 to S_5, the unbiased estimates of M_1 to M_5 become

$$nqr\Sigma_i \ F_i^2/(p-1) + \sigma^2$$
$$npr\Sigma_j \ G_j^2/(q-1) + \sigma^2$$
$$nr \ \Sigma_{ij} I_{ij}^2/(p-1)(q-1) + \sigma^2$$
$$n \ \Sigma_{ijk} J_{k(ij)}^2/pq(r-1) + \sigma^2$$

and σ^2 respectively.

Hence, tests for the existence of any term in the model (3) are made by dividing the appropriate mean square by the Within Cells (or residual) mean square.

To examine the case under Model II, we must first assume that the F's, G's, I's and J's are random samples from normal populations with zero means and variances σ_F^2, σ_G^2, σ_I^2 and σ_J^2. Hence the unbiased estimates of M_1 to M_5 become

$$nqr\,\sigma_F^2 + nr\,\sigma_I^2 + n\,\sigma_J^2 + \sigma^2$$
$$npr\,\sigma_G^2 + nr\,\sigma_I^2 + n\,\sigma_J^2 + \sigma^2$$
$$nr\,\sigma_I^2 + n\,\sigma_J^2 + \sigma^2$$
$$n\,\sigma_J^2 + \sigma^2$$

and σ^2 respectively.

Again, accurate tests for the various terms are again available, but they are not so simple as under Model I. Factor C is tested by the ratio M_4/M_5, the interaction by the ratio M_3/M_4, factor B by the ratio M_2/M_3, and factor A by the ratio M_1/M_3.

Tests for the various effects under the Mixed Model are easily obtained when it is noted that

(i) if factor A has a systematic effect, the term involving I will disappear from the average value of M_2;

(ii) if factor B has a systematic effect, the term involving I will disappear from the average value of M_1;

(iii) if factor C has a systematic effect, the term involving J will disappear from the average value of M_1, M_2 and M_3.

It may be noted in all the cases we have considered, that three important steps are involved in any analysis. Firstly, an appropriate form for the data must be chosen. This will involve deciding whether

the factors involved are crossed or nested. Secondly, the appropriate
partition of the overall sum of squares must be decided upon. The partitioned sums of squares will correspond to the terms which have been
introduced into the assumed form for the data. It will be seen from the
examples we have considered that factors which are nested within
another factor can only occur singly in the analysis of variance table,
but that crossed factors can enter into all possible combinations with
other crossed factors to produce the various interaction sums of
squares which we have derived. They can also form interactions with
any nested factor which has occurred above them in the hierarchy of
factors, because this nested factor must be nested within another factor. Finally, we must work out the average values of the mean squares
which our table has given. This is most important, since it is from
these average values that the various tests are deduced. Once the factors have been classified according as to whether their effects are random or systematic, the working out of these average values is simply a
matter of algebra. However, this may be a tedious business if the
example should contain several factors. It may be somewhat eased if
it is remembered that where a factor has a systematic effect, then any
term in the assumed form for the data which has a dot in place of that
particular suffix can be put equal to zero. As one would expect, this
rule does not apply to suffices which occur inside brackets.

It is useful to remember that the number of degrees of freedom of
interaction mean squares is the product of the degrees of freedom of
the interacting factors. The number of degrees of freedom of a crossed
factor is one less than the number of levels of that factor, and the degrees of freedom of a nested factor is the same multiplied by the number of levels of all of the factors in which it is nested.

A more general example is the case of four factors where B is nested within A and D is nested within A, B and C; A and C being crossed
classifications. The appropriate form for the data in this case is

$$x_{ijkl\alpha} = \mu + A_i + B_{j(i)} + C_k + I_{ik} + J_{jk(i)} + D_{l(ijk)} + \epsilon_{ijkl\alpha}$$

This equation is built up from what we are told about the factors.
The A, B, C and D terms correspond to the main effects of the four
factors, the suffices in brackets indicating that the factor is nested.
The two crossed factors A and C provide an interaction term I_{ik}
and there is also an interaction term of B with C which, of course,
must be nested within A. The partitioning of the sum of squares
would provide seven lines in the analysis of variance table corresponding to the four main effects (two nested), the two interactions and a
"Within Cells" or residual sum of squares.

3.5 Computational note

As before, the various sums of squares which have been quoted should be written in an alternative form before computing begins. Let us consider the summation terms given in the table in section 3.2. The first thing to do is to form all the subsidiary totals of the data: the sum over each cell which is denoted by T_{ijk}, and the sums over the respective factors ($T_{ij.}$, $T_{i \cdot k}$ and $T_{.jk}$). We will also need the sums over the factors taken two at a time ($T_{i..}$, $T_{.j.}$ and $T_{..k}$) as well as the grand total $\Sigma_{ijk\alpha} x_{ijk\alpha} = T....$. Several checks are available for these totals, such as

$$\Sigma_i T_{ij.} = \Sigma_k T_{.jk} = T_{.j.}$$

By completing the squares, we can now write the sums as follows:

$$S = \Sigma_{ijk\alpha}(x_{ijk\alpha} - x....)^2 = \Sigma_{ijk\alpha} x_{ijk\alpha}^2 - \frac{(T...)^2}{N}$$

$$S_1 = nqr\Sigma_i(x_{i...} - x....)^2 = \frac{\Sigma_i T_{i..}^2}{nqr} - \frac{(T...)^2}{N}$$

$$S_4 = nr\Sigma_{ij}(x_{ij..} - x_{i...} - x_{.j..} + x....)^2$$

$$= \frac{\Sigma_{ij} T_{ij.}^2}{nr} - \frac{\Sigma_i T_{i..}^2}{nqr} - \frac{\Sigma_j T_{.j.}^2}{npr} + \frac{(T...)^2}{N}$$

$$S_8 = \Sigma_{ijk\alpha}(x_{ijk\alpha} - x_{ijk.})^2 = \Sigma_{ijk\alpha} x_{ijk\alpha}^2 - \frac{\Sigma_{ijk}(T_{ijk})^2}{n}$$

where the summation is, of course, over all the suffices in the following term. The other main effect and two-factor interactions can be obtained by analogy with S_1 and S_4. The three-factor interaction term can also be expressed in terms of the subsidiary totals, but as this involves eight terms, it is usual to obtain this quantity by subtracting the other sums from S. If it is required as a check on the previous arithmetic then S_7 may be calculated from

$$\frac{\Sigma_{ijk} T_{ijk}^2}{n} - \frac{\Sigma_{ij} T_{ij.}^2}{nr} - \frac{\Sigma_{ik} T_{i \cdot k}^2}{nq} - \frac{\Sigma_{jk} T_{.jk}^2}{np} + \frac{\Sigma_i T_{i..}^2}{nqr}$$

$$+ \frac{\Sigma_j T_{.j.}^2}{npr} + \frac{\Sigma_k T_{..k}^2}{npq} - \frac{(T...)^2}{N}$$

The remarks in section 2.7 regarding the desirability of desk computers for computation of this kind are now more urgent as the complexity of the data has increased.

3.6 Transformation of data

We have frequently assumed in our examples that the parent population is normally distributed. When normality cannot be assumed, the quotients in the analysis of variance table will follow some unknown distribution, and exact tests of significance are not available. To study the effect, on the various tests we have described, of deviations from normality, is long and tedious work. However, it seems to be the general opinion of modern authors that moderate departures from normality can be tolerated, and hence the analysis of variance techniques have a fairly wide application.

One method of making data, which are obviously non-normal, more suitable for analysis of variance work is by means of a transformation. These transformations are used when the variance is a function of the mean, and the object is then to stabilise the variance of the population. Suppose y and x are variates connected by the equation $y = f(x)$ and we wish to choose the function in such a way that the variance of y will be more constant than the variance of x. Assume that x is distributed about a mean of m with a small standard deviation; then to a first approximation we may put

$$y = f(m) + (x - m)f'(m) \qquad (4)$$

Hence the mean value of y is $f(m)$ and the variance of y is

$$\{f'(m)\}^2 \times (\text{variance of } x) \qquad (5)$$

If the variance of x is assumed to be some function of m, say $g(m)$, and the variance of y is to be held constant equal to A, say, then (5) becomes

$$f'(m) = \sqrt{\frac{A}{g(m)}} \quad \text{or} \quad f(m) = \int \sqrt{\frac{A}{g(m)}}\, dm \qquad (6)$$

Hence (6) now provides us with the appropriate form for the transformation. For example, we shall consider the Poisson distribution in which $g(m) = m$; then (6) becomes

$$f(m) = \int \sqrt{\frac{A}{m}}\, dm = 2\sqrt{Am}$$

This suggests that the appropriate transformation to hold the variance more constant will be $y = \sqrt{x}$. This square-root transformation is usually applied to all data which have been counted, as these will probably be Poisson in form.

When we apply analysis of variance techniques to transformed data we have made sure the assumption that the different classes are distributed with the same variance is justified, at least to a close approximation.

A number of other transformations have been proposed, as well as the square-root transformation mentioned above. A summary and bibliography are given by Bartlett (1947).

CHAPTER 4
APPLICATION TO REGRESSION ANALYSIS

4.1 Introduction

It may be noticed that data on which regression analysis is carried out are similar in form to those on which we have been examining analysis of variance techniques. The only difference is that, for regression analysis, it is necessary for the levels of the factors to take a quantitative form, whereas this is not necessary for analysis of variance techniques. In the usual case, certain specific values are chosen for the independent factors, and hence they must be regarded as having a systematic effect. Even when the values of a factor x arise by selection of some other criteria (e.g. if the values of x correspond to the wheat yields of the English counties), the regression equation is calculated assuming these values of x to be fixed. Hence, when we apply analysis of variance techniques to regression analysis, it will only be necessary to consider the analysis under Model I. We shall now develop significance tests for the gradient of the true regression equation.

4.2 Linear regression on one variable

Suppose we are given n pairs of observations (x_i, y_i). It is customary to fit to these data a straight line of the form $y = \alpha + \beta(x - \bar{x})$, where α and β are estimated by $a = \bar{y}$ and $b = \dfrac{\Sigma_i (y_i - \bar{y})(x_i - \bar{x})}{\Sigma_i (x_i - \bar{x})^2}$

It will be assumed that the values of y_i which it is possible to obtain by repeating the experiment for a given value of x_i are normally distributed about a mean of $\alpha + \beta(x_i - \bar{x})$ with variance σ^2. This is equivalent to putting

$$y_i = \alpha + \beta(x_i - \bar{x}) + \epsilon_i \tag{1}$$

where ϵ_i is a normal deviate with zero mean and variance σ^2. Let Y_i be the value of y predicted by the regression equation for a given value of x_i, then

$$Y_i = a + b(x_i - \bar{x}) \tag{2}$$

We may now write $(y_i - \bar{y}) = (Y_i - \bar{y}) + (y_i - Y_i)$

Squaring both sides and summing over i, we get

$$\Sigma_i (y_i - \bar{y})^2 = \Sigma_i (Y_i - \bar{y})^2 + \Sigma_i (y_i - Y_i)^2 \tag{3}$$

or $\quad S = S_1 + S_2$

The cross-product term can be shown to vanish by means of the normal equations used in deriving a and b.

Now S will have $(n-1)$ degrees of freedom since it is computed from the deviations of n observations from their mean. S_2 will have $(n-2)$ degrees of freedom since the terms $(y_i - Y_i)$ are subject to two independent linear constraints. S_1 may be written as $b^2 \Sigma_i (x_i - \bar{x})^2$ on substituting from (2) and writing a for $\bar{y}$. Thus, as the x_i are assumed constant, S_1 is the square of a linear function of normally distributed variables and will have one degree of freedom. These results are shown in the following analysis of variance table:

Source of estimate	Sums of squares	Degrees of freedom	Mean squares
Linear regression	$S_1 = b^2 \Sigma_i (x_i - \bar{x})^2$	1	$M_1 = S_1$
Residual	$S_2 = \Sigma_i (y_i - Y_i)^2$	$n - 2$	$M_2 = \dfrac{S_2}{n-2}$
Total	$S = \Sigma_i (y_i - \bar{y})^2$	$n - 1$	

Under our initial assumptions it is easy to show that as b is a linear function of the y_i's it must be normally distributed with mean β and variance $\dfrac{\sigma^2}{\Sigma_i (x_i - \bar{x})^2}$. Hence b^2 will be an unbiased estimate of $\beta^2 + \dfrac{\sigma^2}{\Sigma_i (x_i - \bar{x})^2}$. Therefore S_1 is an unbiased estimate of $\beta^2 \Sigma_i (x_i - \bar{x})^2 + \sigma^2$. S_2 is, of course, an unbiased estimate of $(n-2)\sigma^2$. Hence M_1 is an unbiased estimate of $\beta^2 \Sigma_i (x_i - \bar{x})^2 + \sigma^2$ and M_2 is an unbiased estimate of σ^2.

If we now make the hypothesis that $\beta = 0$, i.e. that the two variables are unrelated, then M_1 and M_2 are both estimates of σ^2 and we can use the ratio M_1/M_2 to test this hypothesis by using the F tables with $(1, n-2)$ degrees of freedom.

By this partitioning of the sums of squares we have split off from the total variation of the data a term $S_1 = \Sigma_i (Y_i - \bar{y})^2$ which measures the variation arising from the linearity. The term $S_2 = \Sigma_i (y_i - Y_i)^2$ measures the inherent variation of the data about the assumed regression line.

A similar analysis can be carried out for the more general cases of linear regression with more than one variable and of polynomial regression, and these will now be described briefly. They are rarely used, due to the heavy computation involved and limited application.

4.3 Linear regression with more than one variable

Consider the case of linear regression on two variables x_1 and x_2. The data will now consist of n sets of observations (y_i, x_{1i}, x_{2i}). The equation of the regression line is assumed to be

$$y = \alpha + \beta_1(x_1 - \bar{x}_1) + \beta_2(x_2 - \bar{x}_2)$$

It is assumed that each value of y_i is normally distributed with variance σ^2 and so we may write

$$y_i = \alpha + \beta_1(x_{1i} - \bar{x}_1) + \beta_2(x_{2i} - \bar{x}_2) + \epsilon_i \qquad (1)$$

where ϵ_i is the usual error term.

Let Y_i be the value of y predicted from the regression equation, then

$$Y_i = a + b_1(x_{1i} - \bar{x}_1) + b_2(x_{2i} - \bar{x}_2) \qquad (2)$$

where a, b_1 and b_2 are the sample estimates of α, β_1 and β_2

As in section 4.2, we obtain the equation

$$\Sigma_i(y_i - \bar{y})^2 = \Sigma_i(Y_i - \bar{y})^2 + \Sigma_i(y_i - Y_i)^2$$

or

$$S = S_1 + S_2$$

and the following analysis of variance table results:

Source of estimate	Sums of squares	Degrees of freedom	Mean squares
Regression	$S_1 = \Sigma_i \{b_1(x_{1i} - \bar{x}_1) + b_2(x_{2i} - \bar{x}_2)\}^2$	2	$M_1 = \frac{1}{2} S_1$
Residual	$S_2 = \Sigma_i(y_i - Y_i)^2$	$n - 3$	$M_2 = \dfrac{S_2}{n-3}$
Total	$S = \Sigma_i(y_i - \bar{y})^2$	$n - 1$	

Now M_1 is an unbiased estimate of $\frac{1}{2}\Sigma_i\{\beta_i(x_{1i}-\bar{x}_1) + \beta_2(x_{2i}-\bar{x}_2)\}^2 + \sigma^2$

and M_2 " "   "   "   " σ^2

The ratio M_1/M_2 can now be used to test the hypothesis that both β_1 and β_2 are equal to zero. Only in the case when $\Sigma_i(x_{1i}-\bar{x}_1)(x_{2i}-\bar{x}_2)=0$ can the regression sum of squares be split into two parts each with one degree of freedom, giving separate tests for $\beta_1 = 0$ and $\beta_2 = 0$. However, we may artificially split the regression sum of squares by introducing a new variable $x_{2.1}$ in place of x_2 such that $x_{2.1}$ is orthogonal to x_1. Let

$$x_{2.1} = (x_{2i}-\bar{x}_2) - b_{21}(x_{1i}-\bar{x}_1)$$

where

$$b_{21} = \frac{\Sigma_i(x_{2i}-\bar{x}_2)(x_{1i}-\bar{x}_1)}{\Sigma_i(x_{1i}-\bar{x}_1)^2}$$

Then S_1 becomes

$$\Sigma_i\{(b_1 + b_2 b_{21})(x_{1i}-\bar{x}_1) + b_2 x_{2.1}\}^2$$
$$= (b_1 + b_2 b_{21})^2 \Sigma_i(x_{1i}-\bar{x}_1)^2 + b_2^2 \Sigma_i x_{2.1}^2 \qquad (3)$$

Now

$$b_1 = \frac{\Sigma_i X_{1i}(y_i-\bar{y})\Sigma_i X_{2i}^2 - \Sigma_i X_{2i}(y_i-\bar{y})\Sigma_i X_{1i} X_{2i}}{\Sigma_i X_{1i}^2 \Sigma_i X_{2i}^2 - (\Sigma_i X_{1i} X_{2i})^2}$$

and

$$b_2 = \frac{\Sigma_i X_{2i}(y_i-\bar{y})\Sigma_i X_{1i}^2 - \Sigma_i X_{1i}(y_i-\bar{y})\Sigma_i X_{1i} X_{2i}}{\Sigma_i X_{1i}^2 \Sigma_i X_{2i}^2 - (\Sigma_i X_{1i} X_{2i})^2}$$

where $X_{1i} = x_{1i}-\bar{x}_1$ and $X_{2i} = x_{2i}-\bar{x}_2$

Hence $b_1 + b_2 b_{21}$ reduces to $\dfrac{\Sigma_i(x_{1i}-\bar{x}_1)(y_i-\bar{y})}{\Sigma_i(x_{1i}-\bar{x}_1)^2}$ which is the estimate of the regression slope of y on x_1 alone.

From (3), S_1 is now split into two sums of squares each with one degree of freedom, and the first of these sums of squares can be used to test the hypothesis that the regression of y on x_1 alone is zero, while the second can be used to test the hypothesis that $\beta_2 = 0$.

In the case in which k variables are involved the sum of squares can be partitioned into a residual term having $n-k-1$ degrees of freedom and k further terms each having one degree of freedom if a transformation is used to introduce k new variables which are orthogonal. When this has been suitably carried out, the k terms can be used to test the existence of the appropriate regression coefficients. For a more detailed discussion the reader is referred to the standard works on multiple regression.

The problem of fitting the best kth order polynomial $y = a + b_1 x + b_2 x^2 + \ldots + b_k x^k$ to a set of data can now be dealt with by replacing $x, x^2, \ldots, x^k$ by $x_1, x_2, \ldots, x_k$. However, the transformation to a new set of variables can be avoided if, instead of representing the curve by a series of powers of x, it is represented by a series of orthogonal polynomials. The use of orthogonal polynomials in curve fitting is now a standard procedure.

4.4 Example

Before considering a numerical example we must first change the sums of squares required into a form more suitable for computation. In section 4.2 the sum of squares due to the regression S_1 was given as $b^2 \Sigma_i (x_i - \bar{x})^2$. But the gradient of the regression line b is equal to $\Sigma_i (x_i - \bar{x})(y_i - \bar{y}) / \{\Sigma_i (x_i - \bar{x})^2\}$. Hence it is easily seen that S_1 becomes

$$\frac{[\Sigma_i (x_i - \bar{x})(y_i - \bar{y})]^2}{\Sigma_i (x_i - \bar{x})^2} \tag{1}$$

and both summations can be expanded into standard computational forms as follows:

$$\Sigma_i (x_i - \bar{x})(y_i - \bar{y}) = \Sigma_i x_i y_i - \{(\Sigma x_i \Sigma y_i)/n\}$$

$$\Sigma_i (x_i - \bar{x})^2 = \Sigma_i (x_i)^2 - \{(\Sigma x_i)^2/n\}$$

The simplest way of calculating S_2 is by subtracting S_1 from S giving

$$S_2 = \Sigma_i (y_i - \bar{y})^2 - \frac{[\Sigma_i (x_i - \bar{x})(y_i - \bar{y})]^2}{\Sigma_i (x_i - \bar{x})^2} \tag{2}$$

and $\Sigma_i (y_i - \bar{y})^2$ is calculated from $\Sigma y_i^2 - \{(\Sigma y_i)^2/n\}$.

A suitable example concerns the life-testing of electric-light bulbs. Eight bulbs were taken and their life was measured (y) and also a physical measurement (x) on the filament of each bulb. From these data S_1 and S_2 were calculated using formulas (1) and (2) above, and the analysis of variance table on the following page obtained.

The value obtained for the F-ratio is significant, showing that the slope of the regression line connecting x and y is significantly different from zero. In other words, we conclude that the apparent relationship between x and y is real and did not occur by chance.

Source of variation	Sums of squares	Degrees of freedom	Mean squares	F-ratio
Regression	$S_1 = 5752$	1	5752	9.7
Residual	$S_2 = 4163$	7	594.7	
Total	$S = 9915$	8		

CHAPTER 5
SPECIAL TYPES OF ANALYSIS

5.1 Introduction

In order to deal with certain frequently recurring problems in experimental design, modifications of the full analysis of variance procedure become necessary, and it is the more common of these special designs which we shall now consider. The need for these becomes apparent in the case of experiments in which there are one or more important factors affecting the results, and these factors cannot be controlled and measured during the experiment. To ignore them would be to pool their effect with the residual sum of squares, thus artificially inflating the residual mean square and decreasing the chance of finding any significant effects. Clearly the effect of such factors has to be taken into account during the analysis of the experiment.

An example of such a factor would be the variation in fertility of the plots on which crops were being grown. Indeed, it was the science of agriculture which initiated and stimulated the rapid growth of the subject of experimental design. Today, these special designs are used in a wide range of industries such as chemical engineering and electronic components.

Ideally, we should like to be able to control all the factors during an experiment except those which are being investigated. Unfortunately such control is often costly and may be impracticable. The problem is to design an experiment so that known sources of error in the results can be removed during the analysis stage. It should be stressed here that such refinements of analysis can only be carried out when the experiment had been correctly planned, as described below.

5.2 Randomised blocks

The most frequent source of experimental error is the variation between raw materials on which it is necessary to experiment. Suppose that we wish to compare a number of treatments and that the experimental material is available in homogeneous pieces which are large enough to allow all the treatments to be carried out only once per piece. One experimental result per treatment is not sufficient for practical purposes, and yet any repeat of the experiment must take place on a separate piece of material, which cannot be regarded as having the same effect as the first piece. Suppose there are k different treatments to be compared and that n pieces of material have been used, giving kn results

in all. The pieces of material will now be referred to us as "blocks".
The appropriate model to represent these data is

$$x_{ti} = \mu + B_i + T_t + \epsilon_{ti} \quad (t = 1...k, \; i = 1...n) \tag{1}$$

where B_i represents the block effect and T_t the effect of an individual treatment. The two assumptions which are implicit in this model are that (i) there is no interaction between block and treatment and (ii) the error term ϵ_{ti} is assumed to be normally distributed with zero mean and variance σ^2. With these assumptions, this experiment can now be analysed as a conventional two factor analysis with one reading per cell. By comparison with section 2.3 the following analysis of variance table can be built up.

Source of estimate	Sums of squares	Degrees of freedom
Between treatments	$S_t = n \Sigma_t (x_{t.} - x_{..})^2$	$(k - 1)$
Between blocks	$S_B = k \Sigma_i (x_{.i} - x_{..})^2$	$(n - 1)$
Residual	$S_R = \Sigma_{it} (x_{ti} - x_{t.} - x_{.i} + x_{..})^2$	$(n - 1)(k - 1)$
Total	$S = \Sigma_{it} (x_{ti} - x_{..})^2$	$(nk - 1)$

In the usual way we can derive a test for the existence of treatment effects, whether they are random or systematic. Such a test is obtained by dividing the treatment mean square by the residual mean square. It is not necessary to make an assumption regarding the effect of the blocks, although frequently the blocks will have been selected at random from some larger population of possible blocks. It is usually stated also that the treatments should be allocated at random within each block so that any possible systematic variations within the blocks are kept to a minimum.

If the effects of the blocks had been ignored in the analysis, the analysis of variance table would have been the same as above, except that the lines "Between blocks" and "Residual" would have been summed, giving a Residual sum of squares equal to $S_B + S_R$ and with $(nk - k)$ degrees of freedom. In other words, the variation between blocks would have been pooled with the experimental error, thus making a significant variation in treatments harder to detect. By arranging the experiment systematically we have removed this known block effect from the analysis before proceeding to the examination of treatment differences.

When the blocks are large enough, it may be possible to carry out a two-factor analysis on each of them. We denote the results of such an experiment by x_{ijt} where $i = 1 \ldots p$; $j = 1 \ldots q$; $t = 1 \ldots n$; p and q are the number of levels of the two factors respectively; and n is the number of blocks used. The appropriate model can now be written as

$$x_{ijt} = \mu + X_i + Y_j + I_{ij} + B_t + \epsilon_{ijt} \tag{2}$$

where B_t represents the block effect, and the main effects and interaction of the two factors are represented by X, Y and I, respectively. Equation (2) can be regarded as a special case of a three-factor experiment in which only one interaction term is present. Thus the following analysis of variance table can be derived:

Source of estimate	Sums of squares	Degrees of freedom	Mean squares
Factor X	$S_1 = nq \sum_i (x_{i..} - x_{...})^2$	$p - 1$	M_1
Factor Y	$S_2 = np \sum_j (x_{.j.} - x_{...})^2$	$q - 1$	M_2
Interaction	$S_3 = n \sum_{ij} (x_{ij.} - x_{i..} - x_{.j.} + x_{...})^2$	$(p-1)(q-1)$	M_3
Block	$S_B = pq \sum_t (x_{..t} - x_{...})^2$	$n - 1$	M_B
Residual	$S_4 = \sum_{ijt} (x_{ijt} - x_{ij.} - x_{..t} + x_{...})^2$	$(pq-1)(n-1)$	M_4
Total	$S = \sum_{ijt} (x_{ijt} - x_{...})^2$	$npq - 1$	

The expected values of the mean squares can be listed as shown.

	Model 1	Model 2
M_1	$nq \sum_i X_i^2 / (p-1) + \sigma^2$	$nq \sigma_X^2 + n \sigma_I^2 + \sigma^2$
M_2	$np \sum_j Y_j^2 / (q-1) + \sigma^2$	$np \sigma_Y^2 + n \sigma_I^2 + \sigma^2$
M_3	$\dfrac{n}{(p-1)(q-1)} \sum_{ij} I_{ij}^2 + \sigma^2$	$n \sigma_I^2 + \sigma^2$
M_B	$\dfrac{n}{(n-1)} \sum_t B_t^2 + \sigma^2$	$pq \sigma_B^2 + \sigma^2$
M_4	σ^2	σ^2

As for the general two-factor analysis, tests for the two main effects are obtained by dividing M_1 and M_2 by the residual mean square

under Model 1 and by the interaction mean square under Model 2. The
existence of the interaction is tested for in both cases by dividing by
M_4. Similarly the analysis follows an exact parallel with the general
case when the two factors form a Mixed Model.

The analysis has again had the effect of removing the unknown
block effects by splitting off from the residual term a sum of squares
S_B which measures this unknown effect. This leaves the two factors to
be compared against the true experimental error and not an inflated estimate of it. A special case of the two-factor randomised block design
appears in section 5.5 under the title of "Split Plot Designs".

5.3 Latin squares

If the blocks of raw material in the previous paragraph are not
large enough to allow all the treatments to be carried out on each block,
a new approach becomes necessary. It will be assumed that the raw
material available for the experiment is only homogeneous in so far as
to enable a single treatment to be carried out. Further, we shall assume
that this material is available in some planar form, e.g. as a sheet of
paper, a piece of leather, or a plot of ground. It is our intention to
eliminate from this plane as much of the natural variation as possible.
We shall, in the analysis, remove the variation corresponding to two
directions at right angles as the best approximation we can make to
this ideal.

A suitable arrangement for the layout of the treatments is illustra-
ted in the diagrams below. A square whose side is equal to the number
of treatments to be tested is drawn and in each cell of the square a
letter is placed in such a manner that each letter appears once and only

A	B
B	A

C	B	A
A	C	B
B	A	C

D	A	C	B
A	B	D	C
B	C	A	D
C	D	B	A

once in each row and column. The diagrams illustrate a 2×2, 3×3
and a 4×4 Latin square. The square may be thought of as representing the plane of experimental material, and the treatments can be allocated to the squares according to the letters which we have entered in
the cells. For example, if we compare four treatments we may choose
a Latin square as shown above. Then one treatment would be given to
all those areas in which we have written A, another to those areas
where there is a B, and so on. In this way, each treatment appears
once in each row and column, thus enabling the treatment effects to be

separated from row and column effects. It may be noted that the use of such a design makes it necessary to do n^2 experiments when comparing n treatments.

Let us denote the experimental results by x_{ij} where i and j denote the row and column for that particular reading ($i,j = 1 \ldots n$). Then the appropriate model to assume is

$$x_{ij} = \mu + \alpha_i + \beta_j + \gamma_k + \epsilon_{ij} \qquad (1)$$

where α_i is the effect of the ith row
β_j is the effect of the jth column
and γ_k is the effect of the kth treatment.

The subscript k will depend on i and j, the actual value being determined by the particular square being used. The major assumption made by the model in (1) is that the interaction terms are assumed to be zero. We are, in fact, carrying out a 3-factor experiment (row, column and treatment) with only n^2 observations instead of the minimum n^3 as indicated in Chapter 3. Hence we are not able to test for these interactions.

As before, an analysis of variance table can be drawn up as shown:

Source of estimate	Sums of squares	Degrees of freedom
Rows	$n \Sigma_i (x_{i.} - x_{..})^2$	$n - 1$
Columns	$n \Sigma_j (x_{.j} - x_{..})^2$	$n - 1$
Treatments	$n \Sigma_k (x_{(k)} - x_{..})^2$	$n - 1$
Residual	$\Sigma_{ij} (x_{ij} - x_{i.} - x_{.j} - x_{(k)} + 2x_{..})^2$	$(n-1)(n-2)$
Total	$\Sigma_{ij} (x_{ij} - x_{..})^2$	$n^2 - 1$

where $x_{(k)}$ represents the mean value corresponding to the kth treatment. As there are no interaction terms, the test for treatment differences is made by dividing the treatment mean square by the residual mean square, whichever model is assumed for the data.

If a Latin square design had not been used, the observations would have given rise to a single-factor analysis of variance table in which the rows, columns and residuals above would have been pooled into one large residual sum of squares, making it more difficult to obtain a significant effect for treatments, should one exist. Thus we have again removed some, at least, of the variation of the experimental material from the analysis before the treatment test is carried out.

This type of design can be generalised in two ways. Firstly, if the raw material does not appear in planar form and there is evidence that there is a variation in quality in three dimensions, then the appropriate design is a Latin cube. In this, each letter appears only once in each row and column and file of the cube, and then the letters are randomly allocated to treatments as before. Thus, variation in raw material in three mutually perpendicular directions can be removed. The second generalisation occurs when it is desired to examine two different factors on the same piece of material. This is done by superimposing two different Latin squares in such a way that a given letter of one square occurs with each individual letter of the second square. Such designs are often referred to as Graeco-Latin squares, since the letters in the second square are usually denoted by Greek letters to avoid confusion. For further details the reader should consult any textbook on the design of experiments, e.g. Davies (1960).

5.4 Balanced incomplete blocks

In an ordinary randomised block design, the block must be large enough to take one complete set of treatments. This is frequently not possible, although the block may allow more than one experiment. Suppose the homogeneous blocks will allow the execution of four experiments, and we wish to compare five treatments. If one of the experiments is carried out on a separate block, then any comparisons will be invalidated due to the differences between blocks. One way to avoid this would be to include in every block one of the treatments as a "standard"; then every treatment can be compared with the standard, and all comparisons made. However, this is not so efficient as the following design, in which the five treatments are replicated four times. To do this five blocks of material have been used.

BLOCK NO.

1	2	3	4	5
A	A	A	A	B
B	B	B	C	C
C	C	D	D	D
D	E	E	E	E

The important point of such a design is that every pair of treatments occurs together in a block the same number of times. It is for this reason that the design is referred to as "balanced". Balanced Incomplete Block designs belong to the wider class of incomplete block designs which are outside the scope of this book.

To generalise from this special case, let b = number of blocks, k = number of treatments each block will hold, t = number of treatments, and r = number of replications of each treatment; then it is easily seen that in a balanced design we must have

$$bk = tr \qquad (1)$$

Now any treatment occurs in just r blocks. Then in these blocks there are a total of $r(k-1)$ other positions in which we can allocate the remaining $(t-1)$ treatments. Thus any other treatment will occur with the first treatment $r(k-1)/(t-1)$ times. That is, a given pair of treatments will occur

$$\lambda = \frac{r(k-1)}{(t-1)} \quad \text{times} \qquad (2)$$

As before, we assume no interactions between the treatments and blocks, and the appropriate model for the data becomes

$$x_{ij} = \mu + T_i + B_j + \epsilon_{ij} \qquad (i = 1\ldots t,\ j = 1\ldots b) \qquad (3)$$

with T_i representing the treatment effect and B_j representing the block effect. The error term is assumed to be normally distributed with zero mean and variance σ^2. As before, the following analysis of variance table can be built up where $N = tr$. It should be remembered that all combinations of i and j are not permitted.

Source of estimate	Sums of squares	Degrees of freedom
Between treatments	$S_1 = r\sum_i (x_{i.} - x_{..})^2$	$t - 1$
Between blocks	$S_2 = k\sum_j (x_{.j} - x_{..})^2$	$b - 1$
Residual	$S_3 = \sum_{ij} (x_{ij} - x_{i.} - x_{.j} + x_{..})^2$	$N - b - t + 1$
Total	$S = \sum_{ij} (x_{ij} - x_{..})^2$	$N - 1$

Again it can be shown that whatever type of factor we assume for the treatments and blocks, there is a valid test for treatment differences obtained by dividing the treatment mean square by the residual mean square. Thus it may be stated that this design is a direct extension of the Randomised Blocks of section 5.2, although greater care is needed in the calculation of the sums of squares, and the student should study the example given in section 5.6.

5.5 Split plot designs

In this design, we are particularly concerned with a special case of the two-factor randomised block design. Here we want to obtain accurate information about one factor and also about the interaction between the two factors, the second factor being of no immediate concern to the experimenter. Suppose we wish to assess the effect of factor A and the interaction of factor A with factor C. In this case the p treatments of factor C would be arranged in a randomised block design of r blocks as described. Each of the pr plots is now divided into q subplots so that the q treatments of factor A can be allocated at random over each plot. The advantage of this type of design is that because no attempt is to be made to get an accurate assessment of factor C, larger plots can be used for allocation of the first p treatments, irrespective of the variation within the blocks.

We assume the following model for the data:

$$x_{ijk} = \mu + A_i + C_j + B_k + (BC)_{jk} + (AC)_{ij} + \epsilon_{ijk}$$
$$(i = 1 \ldots q,\ j = 1 \ldots p,\ k = 1 \ldots r)$$

where A_i and C_j are the main effects and B_k is the block effect as before. Notice that we must now assume the possibility of an interaction between the block and factor C, since we are not now stipulating that the blocks should be regarded as homogeneous. What we are assuming is that the pr plots within the blocks are themselves sufficiently homogeneous for a complete comparison of the treatments of factor A. Hence there is no interaction between blocks and factor A written into the formula above.

The analysis of variance table is built up in exactly the same manner as before:

Source of estimate	Sums of squares	Degrees of freedom
Factor A	$S_1 = pr \sum_i (x_{i..} - x_{...})^2$	$q - 1$
Factor C	$S_2 = qr \sum_j (x_{.j.} - x_{...})^2$	$p - 1$
Block	$S_3 = pq \sum_k (x_{..k} - x_{...})^2$	$r - 1$
$A \times C$ interaction	$S_4 = r \sum_{ij} (x_{ij.} - x_{i..} - x_{.j.} + x_{...})^2$	$(p-1)(q-1)$
$C \times$ block interaction	$S_5 = p \sum_{ik} (x_{i.k} - x_{i..} - x_{..k} + x_{...})^2$	$(q-1)(r-1)$
Residual	$S_6 = \sum_{ijk} (x_{ijk} - x_{i.k} - x_{ij.} + x_{i..})^2$	$q(p-1)(r-1)$
Total	$S = \sum_{ijk} (x_{ijk} - x_{...})^2$	$pqr - 1$

Tests for the existence of an effect due to factor A and the $A \times C$ interaction can be made by dividing the appropriate mean squares by the residual mean square. A test for the existence of an effect due to factor C is available by dividing the mean square due to factor C by the $C \times$ Block interaction. In this sense, we may regard factor C and the blocks as forming a separate two-factor experiment. Because this test is based on smaller degrees of freedom than the previous ones, it will be less sensitive, and it is for this reason that a Split Plot design is normally used only when information about the second factor is of relative unimportance. For a more complete description of this type of design, the reader should consult, for example, Cochran and Cox (1957).

5.6 Examples

I – *Randomised blocks*

The following data refer to three methods of hardening steel, which are referred to as A, B, and C. A long strip of steel was cut into 18 pieces and the treatments allocated at random, forming six complete blocks. The actual layout of the treatments and data is shown below:

A	B	C	C	B	A	A	C	B
813	647	713	814	759	795	705	652	598

A	C	B	B	A	C	A	B	C
774	617	559	580	687	539	581	480	437

Evaluating the analysis of variance table by means of the computational methods outlined in previous chapters, we get:

Source of variation	Sums of squares	Degrees of freedom	Mean squares	Ratio of mean squares
Treatments	49884·1	2	24942·1	13·3
Blocks	149700·4	5	29940·1	
Residual	18724·6	10	1872·5	
Total	218309·1	17		

As the ratio of mean squares is 13·3 and the 1 per cent value of the variance ratio with (2, 10) degrees of freedom is 7·57, there is a very strong indication of real treatment differences. If the effect of the block, which in this case corresponds to position along the strip, had been ignored, the results form a single-factor experiment with six replicates, giving the following analysis of variance table.

Source of variation	Sums of squares	Degrees of freedom	Mean squares	Ratio of mean squares
Treatments	49884·1	2	24942·1	2·22
Residual	168425	15	11228	
Total	218309·1	17		

Here the ratio of mean squares is not significant, and the conclusion would be that the treatments show no significant differences. We have lost the significant result because we have now got an artificially inflated residual sum of squares due to the variation in the experimental material.

II – *Latin squares*

The following experiment was designed to measure the effect of preconditioning on the rate of abrasion of leather samples. Small samples of leather were cut from a square pattern and allocated treatments according to the letters in a Latin square. As there are six treatments, a 6×6 Latin square was used:

C 7·38	D 5·39	F 5·03	B 5·50	E 5·01	A 6·79
B 7·15	A 8·16	E 4·96	D 5·78	C 6·24	F 5·06
D 6·75	F 5·64	C 6·34	E 5·31	A 7·81	B 8·05
A 8·05	C 6·45	B 6·31	F 5·46	D 6·05	E 5·51
F 5·65	E 5·44	A 7·27	C 6·54	B 7·03	D 5·96
E 6·00	B 6·55	D 5·93	A 8·02	F 5·80	C 6·61

The appropriate sums of squares are computed as before and the analysis of variance table becomes:

Source of variation	Sums of squares	Degrees of freedom	Mean squares	Ratio of mean squares
Rows	2·1897	5	0·438	
Columns	2·5742	5	0·515	
Treatments	23·5300	5	4·706	27·2
Residual	3·4888	20	0·175	
Total	31·7827	35		

Hence there is a very significant effect due to treatments. By use of the Latin square design we have split off a good proportion of the residual sum of squares. Hence the accuracy of the estimates of the treatment effects has been increased.

III – *Balanced incomplete blocks*

The data represent the effects of five different tests of ten bales of rubber. Each sample of rubber was only large enough for two tests to be carried out. Therefore, using the notation of section 5.4, we have:

$$b = 10, \quad k = 2, \quad t = 5, \quad r = 4, \quad N = tr = 20$$

	Treatment	1	2	3	4	5	Total
	1	35	16	–	–	–	51
	2	20	–	10	–	–	30
	3	13	–	–	26	–	39
Bale	4	25	–	–	–	21	46
	5	–	16	5	–	–	21
number	6	–	21	–	24	–	45
	7	–	27	–	–	16	43
	8	–	–	20	37	–	57
	9	–	–	15	–	20	35
	10	–	–	–	31	17	48
Total		93	80	50	118	74	415

The following analysis of variance table can then be calculated:

Source of variation	Sums of squares	Degrees of freedom	Mean squares	Ratio of mean squares
Treatments	503·4	4	125·85	3·77
Bales	504·25	9	56·02	
Residual	200·1	6	33·35	
Total	1207·75	19		

As the ratio of mean squares is 3·77 and the 5 per cent value of the variance ratio with (4, 6) degrees of freedom is 4·5, there is no evidence of any significant difference between treatments in this experiment.

IV – *Split plot*

Suppose we wish to compare the efficiencies of four different tests, by examining their results (factor A), and four different machines (factor C) are used to supply test material. The output of one machine will form a whole plot, and each plot is divided into four subplots for the four tests. The experiment was repeated four times and the following 64 results obtained:

	Test 1	Test 2	Test 3	Test 4
Machine 1	81·8	46·2	78·6	77·4
	72·2	51·6	70·9	73·6
	72·9	53·6	69·8	70·3
	74·6	57·0	69·6	72·3
Machine 2	74·1	49·1	72·0	66·1
	76·2	53·8	71·8	65·5
	71·1	43·7	67·6	66·2
	67·8	58·8	60·6	60·6
Machine 3	68·4	54·5	72·0	70·6
	68·2	47·6	76·7	75·4
	67·1	46·4	70·7	66·2
	65·6	53·3	65·6	69·2
Machine 4	71·5	50·9	76·4	75·1
	70·4	65·0	75·8	75·8
	72·5	54·9	67·6	75·2
	67·8	50·2	65·6	63·3

and the analysis of variance is as follows.

Source of variation	Sums of squares	Degrees of freedom	Mean squares	Ratio of mean squares
Factor A	4107·39	3	1369·13	81
Factor C	194·56	3	64·85	3·69
Block	223·81	3	74·60	
$A \times C$ interaction	221·74	9	24·64	
$C \times$ block interaction	158·25	9	17·58	
Residual	608·47	36	16·90	
Total	5514·22	63		

There is a highly significant variation between the tests ($F = 1{,}369\cdot 13/16\cdot 90$ with 3 and 36 degrees of freedom) but not between the output of the four machines ($F = 64\cdot 85/17\cdot 58 = 3\cdot 69$ is not significant). Also there is no evidence of the existence of any interaction between machines and tests.

CHAPTER 6
ANALYSIS OF COVARIANCE

6.1 Introduction
It sometimes happens that during the carrying out of a carefully designed experiment there is an uncontrollable variable which varies between the runs of the experiment. As well as measuring the individual results, we must measure the value of this intruding variable at the time of each run. The data thus consist of pairs of observations to be processed. Before a conventional analysis of variance can be performed, the effect of the second variable must be removed by a method analogous to the regression analysis described in Chapter 4. The technique of dealing with these pairs of observations is known as analysis of covariance.

Covariance analysis has been described as "the technique of testing for homogeneity in problems dealing with two or more correlated variables" (Wishart and Sanders, 1936). The algebra will be found to be exactly similar to that used in analysis of variance. Essentially, we shall be partitioning the sum of products of the deviations of the variates from their means into components associated with different factors.

6.2 Mathematical derivation for one-way classification
Suppose that the data consist of N pairs of corresponding values of the two variables, x and y, and that these fall into p different groups according to the levels of the factor under investigation. Let (x_{ij}, y_{ij}) be the jth pair of values in the ith group ($i = 1...p$). We shall allow different numbers of observations in each category by putting n_i to be the number of pairs in the ith group. In the notation of previous chapters it is obvious that

$$x_{ij} - x_{..} = (x_{ij} - x_{i.}) + (x_{i.} - x_{..}) \qquad (1)$$
$$y_{ij} - y_{..} = (y_{ij} - y_{i.}) + (y_{i.} - y_{..}) \qquad (2)$$

By multiplying these equations and summing over all pairs of observations we get

$$\Sigma_{ij}(x_{ij} - x_{..})(y_{ij} - y_{..}) = \Sigma_{ij}(x_{ij} - x_{i.})(y_{ij} - y_{i.})$$
$$+ \Sigma_i n_i(x_{i.} - x_{..})(y_{i.} - y_{..}) \qquad (3)$$

since the other two terms on the right-hand side are found to be identically zero. The terms in equation (3) can be labelled if we rewrite the equation $S = S_1 + S_2$.

If we now assume that the N pairs of values form a random sample from a homogeneous population with covariance μ_{11}, the expected value of S becomes $(N-1)\mu_{11}$. In other words, if we assume that the factor being investigated has no effect, then S can be used as an estimate of $(N-1)\mu_{11}$. By restricting attention to the ith group, it is clear that the expected value of $\Sigma_j (x_{ij} - x_{i.})(y_{ij} - y_{i.})$ is $(n_i - 1)\mu_{11}$. Hence the expected value of S_1 will be $\Sigma_i (n_i - 1)\mu_{11}$ which is equal to $(N-p)\mu_{11}$. From (3) we conclude that the expected value of S_2 is $(p-1)\mu_{11}$. Thus, if our assumption is correct, each of the terms of equation (3) could be used to estimate the unknown covariance with degrees of freedom $(N-1)$, $(N-p)$ and $(p-1)$ respectively.

By squaring equations (1) and (2) and summing over all observations we get

$$\Sigma_{ij}(x_{ij} - x_{..})^2 = \Sigma_{ij}(x_{ij} - x_{i.})^2 + \Sigma_i n_i (x_{i.} - x_{..})^2$$
$$\Sigma_{ij}(y_{ij} - y_{..})^2 = \Sigma_{ij}(y_{ij} - y_{i.})^2 + \Sigma_i n_i (y_{i.} - y_{..})^2 \quad (4)$$

The six terms above can be labelled as before by rewriting equation (4) as

$$A = A_1 + A_2$$
$$B = B_1 + B_2$$

and each of these terms could be used to estimate the variance of the x or y variable under our assumption that the factor under investigation has no effect. It may be useful to summarize our notation at this point:

Source of variation	Degrees of freedom	Sums of squares	Sums of products
Between groups	$p - 1$	A_2, B_2	S_2
Within groups	$N - p$	A_1, B_1	S_1
Total	$N - 1$	A, B	S

To each line on this table we could apply the results of equations (1) and (2) of section 4.4. A regression line could be calculated and the sum of squares partitioned. For example, the last line of the table gives a sum of squares due to the regression of S^2/A with 1 degree of freedom and a sum of squares due to the deviation about the regression line of $(B - S^2/A)$ with $(N - 2)$ degrees of freedom.

The first test to be carried out is to see if there is evidence of any regression between the x and y variables, and this can be done on the second row of the table since it is here that the effect of the factor under test has been removed. The significance of the regression is found by comparing S_1^2/A_1, having 1 degree of freedom, with $(B_1 - S_1^2/A_1)$ having $(N - p - 1)$ degrees of freedom, by means of the variance ratio distribution. If no significance is found, a conventional analysis of variance is permissible, ignoring the values of y. However, if the test yields significance, we must correct for this association before testing for the significance of the factor under experiment.

This second test can be performed if we find a "between groups" sum of squares which has been corrected for the effect of regression, to compare with the "within group" corrected sum of squares $(B_1 - S_1^2/A_1)$. The sum of squares which is used is in fact $(B - S^2/A) - (B_1 - S_1^2/A_1)$; this is compared with $(B_1 - S_1^2/A_1)$ by means of the variance ratio test with $(p - 1, N - p - 1)$ degrees of freedom. At first sight, it might appear that $(B_2 - S_2^2/A_2)$ would have been satisfactory as a corrected "between groups" sum of squares. This is not so, since this expression has been corrected for the regression between groups, while $(B_1 - S_1^2/A_1)$ has been corrected for the regression "within groups". The slopes of these two regression lines may not be the same and cannot be assumed so without making use of prior knowledge of the experimental background, or making some further analysis of the data.

6.3 Two-way classification

Consider the two-factor analysis of variance in which each experiment yields measurements on two related variables, x and y. As before, we shall assume that there are N pairs of observations x_{ij} and y_{ij} and that these are classified according to the two factors. If we allow only one pair of observations for each combination of levels of the factors, we may write $i = 1...p$, $j = 1...q$, where p and q represent the number of levels of each factor. It follows that $N = pq$. The algebra again follows closely on that of the two-way analysis of variance. We may start from the two identities

$$(x_{ij} - x_{..}) = (x_{i.} - x_{..}) + (x_{.j} - x_{..}) + (x_{ij} - x_{i.} - x_{.j} + x_{..}) \quad (1)$$

$$(y_{ij} - y_{..}) = (y_{i.} - y_{..}) + (y_{.j} - y_{..}) + (y_{ij} - y_{i.} - y_{.j} + y_{..}) \quad (2)$$

By multiplying these equations together and summing over all observations, we get

$$\Sigma_{ij}(x_{ij}-x_{..})(y_{ij}-y_{..}) =$$
$$= \Sigma_i q(x_{i.}-x_{..})(y_{i.}-y_{..}) + \Sigma_j p(x_{.j}-x_{..})(y_{.j}-y_{..})$$
$$+ \Sigma_{ij}(x_{ij}-x_{i.}-x_{.j}+x_{..})(y_{ij}-y_{i.}-y_{.j}+y_{..}) \qquad (3)$$
or
$$S = S_1 + S_2 + S_3 \qquad (4)$$

As in section 6.2, all the terms in (3) could be used to provide estimates of the covariance of the two variables, assuming that the two factors have no influence on the results, so that the N pairs of values form a random sample from a homogeneous population. Similarly, by squaring equations (1) and (2) we can obtain similar partitions of the sums of squares $\Sigma_{ij}(x_{ij}-x_{..})^2$ and $\Sigma_{ij}(y_{ij}-y_{..})^2$. Denoting these by
$$A = A_1 + A_2 + A_3$$
$$B = B_1 + B_2 + B_3$$
where the subscripts refer to corresponding terms in (8), we can summarize the partitioning of the sums of squares and cross-products as follows:

Source of variation	Degrees of freedom	Sums of squares	Sums of products
First factor	$p-1$	$A_1, \ B_1$	S_1
Second factor	$q-1$	$A_2, \ B_2$	S_2
Error	$(p-1)(q-1)$	$A_3, \ B_3$	S_3
Total	$N-1$	$A, \ B$	S

As there is only one observation for each combination of levels, it is necessary to assume that there is no interaction between the factors. Thus we may refer to the third line as due to error alone. Having satisfactorily partitioned the sums of squares and products, we can again apply the results of Chapter 4 to each line of this table. Firstly, we must test to see if there is any regression between the two variables x and y, and this is done from the third line of the above table, since it is here that the effects of the two factors have been removed. The significance of the regression is found by comparing S_3^2/A_3, having 1 degree of freedom, with $(B_3 - S_3^2/A_3)$ having $(N-p-q)$ degrees of freedom, by means of the variance ratio test. If a significant result is obtained, this association must be corrected for before testing for significant effects due to the two factors being examined.

After correcting for regression, the "error" sum of squares becomes $(B_3 - S_3^2/A_3)$. The corrected sum of squares for the first factor can be found by considering a reduced version of the above table.

Source of variation	Degrees of freedom	Sums of squares	Sums of products
First factor	$p - 1$	$A_1, \ B_1$	S
Error	$(p-1)(q-1)$	$A_3, \ B_3$	S_3
Total	$N - q$	$A_1 + A_3, \ B_1 + B_3$	$S_1 + S_3$

Since the gradient of the regression due to the first factor may differ from the gradient of the regression measured by the error term, it follows that the appropriate corrected sum of squares for the first factor becomes

$$\left\{ (B_1 + B_3) - \frac{(S_1 + S_3)^2}{(A_1 + A_3)} \right\} - (B_3 - S_3^2/A_3)$$

A test for the effect of the first factor is obtained by comparing this with $(B_3 - S_3^2/A_3)$ by means of the variance ratio test with $(p - 1, N - p - q)$ degrees of freedom. A test for the effect of the second factor can be built up similarly.

If we had included n observations for every combination of levels of the two factors we would have been able to include in our initial partitioning of the sums of squares and products a line corresponding to the interaction of the two factors. A test for the significance of the interaction can be built up exactly as for the significance of the factors themselves.

It is important to note that after correction for the regression the procedure used is that of analysis of variance. Consequently all the conditions in the first three chapters regarding the various models involving random and systematic factors are applicable here. In particular, testing of the main effects should be made against either the interaction or the error line in the table, depending on whether the factors are regarded as random or systematic.

From examination of this and the previous sections it should now be clear how the associated variable or co-variable in a conventional analysis of variance experiment can be allowed for. The sums of squares and products are partitioned as for analysis of variance, a test of the existence of regression is made, and if found, this regression must be removed before tests can be made on the significance of the factors and their interactions. The same procedure is followed however many factors are involved, care being taken in each case to ensure that the main effects and interactions are tested against the appropriate corrected sum of squares, depending on the nature of the factors involved.

6.4 Example

In a two-factor experiment in an oil refinery, it was found that the feed gravity changed from experiment to experiment. It was therefore necessary to discover if this gravity change (y) affected the results of the experiments, namely naphtha yield (x). Accordingly, for each combination of the levels of the two factors A and B, a pair of results was measured, x and y. There were four levels of factor A and three of factor B, the experiment being carried out twice to give a total of 24 pairs of observations. Both the factors are assumed to have systematic effects, so that a Model 1 analysis will be appropriate.

As for sums of squares, the sums of cross-products are most easily found by expanding the algebraic expressions and evaluating the separate terms. For example

$$\Sigma_{ij}(x_{ij} - x_{..})(y_{ij} - y_{..}) = \Sigma_{ij} x_{ij} y_{ij} - pq x_{..} y_{..}$$

and the right-hand side of the above is computed. In this way the following table of sums of squares and products was built up:

Source of variation	Degrees of freedom	Sums of squares	Sums of products
Factor A	3	52, 61	−28
Factor B	2	16, 35	− 4
Interaction	6	4, 7	− 5
Error	12	26, 41	−26
Total	23	98, 144	− 63

The first test is always to see if there is any regression between x and y, because if there is not, it is permissible to analyse the x-readings by analysis of variance. This test is always made on the error line in the above table by forming a separate sub-table in order to apply the variance-ratio test.

ERROR

Source of variation	Degrees of freedom	Sums of squares	Mean squares	F-ratio
Regression	1	26	26	18·6
Remainder	11	15	1·4	
Total	12	41		

The sum of squares for the y variable is partitioned as described earlier, and it is easily seen that the value of F is significant, showing that there is a strong regression between x and y; hence the sums of squares must be corrected for this.

As this is a Model 1 analysis both the main effects and the interaction are tested against the error terms. We shall begin with the interaction term. A separate sub-table is formed from the two lines "interaction" and "error" in the first table.

Source of variation	Degrees of freedom	Sums of squares	Sums of products
Interaction	6	4, 7	− 5
Error	12	26, 41	− 26
Total	18	30, 48	− 31

The corrected sums of squares for the error and the total rows can now be formed as described, and the corresponding degrees of freedom reduced by one in each case. A similar process could be carried out for the interaction row, giving a value with only five degrees of freedom. It is more desirable to obtain a corrected sum of squares for the interaction by differencing the error and total rows, since this will give a value with six degrees of freedom. These values are summarized below:

Source of variation	Corrected sums of squares	Appropriate degrees of freedom	Mean squares
Interaction	1*	6*	0·167
Error	15	11	1·363
Total	16	17	

(* See text above)

A non-significant result for the interaction effect is obtained.

Exactly the same procedure is followed to test the significance of the main effects. We shall illustrate it by testing the significance of factor A only. A separate sub-table is formed:

Source of variation	Degrees of freedom	Sums of squares	Sums of products
Factor A	3	52, 61	−28
Error	12	26, 41	−26
Total	15	78, 102	−54

As for the interaction test, the corrected sums of squares for the error and total rows are calculated using the formulae given in the text, and their corresponding degrees of freedom reduced by one. In order to maintain the three degrees of freedom for Factor A, the corrected sum of squares for this row is obtained by differencing the two calculated values as shown below:

Source of variation	Corrected sums of squares	Appropriate degrees of freedom	Mean squares
Factor A	50*	3*	16·7
Error	15	11	1·363
Total	65	14	

(* See text above)

The ratio of the two mean squares gives a value of 12·2, which is larger than the 5 per cent F-value with (3, 11) degrees of freedom. Hence there is evidence of an effect due to factor A on the results of the experiment.

CHAPTER 7

TWO-LEVEL EXPERIMENTS

7.1 Introduction

Experiments in which the factors have only a small number of levels occur frequently. When a new process is being investigated at the pilot plant stage it is often only important to decide which factors have any effects at all. Consequently, by allocating only two or three levels to each factor, a rapid preliminary investigation can be carried out. Also, it may be desirable to keep the number of levels small for reasons of expense, since to investigate fully a set of factors with $n_1, n_2, n_3 \ldots$ levels requires $n_1 \times n_2 \times n_3 \times \ldots$ individual experiments. Because of their popularity, special short-cut methods of computation and special designs have been developed for these limited experiments. It is the classical pilot experiment with only two levels for each factor that we shall be dealing with in this chapter, although we shall indicate later how the methods described can be extended to include factors with three or more levels.

7.2 Notation

A three-factor experiment with each factor having two levels is known as a 2^3 design, since this is the number of experiments to be carried out. If there are n such factors then it is known as a 2^n design. Possibly the easiest way to visualise a 2^3 experiment is to consider the three factors A, B and C acting along three mutually perpendicular directions; then the separate experiments correspond to the vertices of a cube, where the subscripts 1 and 2 refer to the first and second levels of the factors. Thus $A_2 B_1 C_2$ means that A and C are held at their second levels whilst B is held at the first level. It may be noted that the results on the top of the cube are the same as those on the bottom, except that factor A has changed from the lower to the upper level, which agrees with our definition of factor A acting along the vertical axis. Thus the effect of factor A can readily be measured as being the difference between the totals of the experiments on the upper and lower faces of the cube, which equals

$$(A_2 B_1 C_1 + A_2 B_2 C_1 + A_2 B_2 C_2 + A_2 B_1 C_2) - (A_1 B_1 C_1 + A_1 B_2 C_1 + A_1 B_2 C_2 + A_1 B_1 C_2)$$

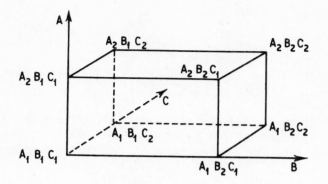

As this is a cumbersome notation, we shall designate by a, b and c the presence of A, B and C at their second levels. Thus ab will mean that A and B are at their second levels and C is at the first. When all the factors are at their lower level, we shall write (1). In this way the effect of factor A becomes

$$(a + ab + abc + ac) - ((1) + b + bc + c) \tag{1}$$

To estimate the interaction between A and B we must first hold B at the lower level. An estimate of A then becomes $(a + ac) - ((1) + c)$. With B at the second level the estimate of A is $(ab + abc) - (b + bc)$. Thus the $A \times B$ interaction may be measured by the difference between these quantities

$$(ab + abc + (1) + c) - (a + ac + b + bc) \tag{2}$$

It will be convenient to denote by A and AB the totals for each effect. Thus we may write

$$\left. \begin{array}{l} A = (a + ab + abc + ac) - ((1) + b + bc + c) \\ AB = (ab + abc + (1) + c) - (a + ac + b + bc) \end{array} \right\} \tag{3}$$

The second-order interaction ABC can now be estimated. For each of the levels of C, the total $A \times B$ interaction is given by

$$(abc + c) - (ac + bc) \quad \text{and} \quad (ab + (1) - (a + b))$$

Hence the total effect due to the second-order interaction is the difference between these quantities

$$ABC = (abc + a + b + c) - (ab + ac + bc + (1)) \tag{4}$$

For the remaining main effects and interactions, equations can be built up which are analogous to (3) and (4). Using our previous notation, these become

$$\left.\begin{array}{rl} B &= abc + ab + bc + b - ac - a - c - (1) \\ C &= abc + ac + bc + c - ab - a - b - (1) \\ AC &= abc + ac + b + (1) - ab - bc - a - c \\ BC &= abc + bc + a + (1) - ab - ac - b - c \end{array}\right\} \quad (5)$$

It is interesting to note that equations (3), (4) and (5) can be represented symbolically by means of the mnemonic

$$\begin{align*} A &= (a-1)(b+1)(c+1) \\ B &= (a+1)(b-1)(c+1) \\ C &= (a+1)(b+1)(c-1) \\ AB &= (a-1)(b-1)(c+1) \text{ etc.,} \end{align*}$$

where a negative sign appears in every bracket corresponding to a term on the left-hand side of the equation. It is important to stress that this last set of equations has no physical significance and serves only to provide a quick way of arriving at the expansions given earlier. This shorthand notation can be extended to include any number of factors. To complete the notation we shall introduce the symbol I corresponding to the sum of all the observations. It is easily seen that $I = (a+1)(b+1)(c+1)$.

Having obtained estimates for the main effects and interactions, these can readily be converted into the appropriate sums of squares for analysis of variance work. The procedure is to square each of the values given by (3), (4) and (5) and divide the results by 8. For the general 2^n experiment, the division would be 2^n which is the number of individual results obtained. If the design is repeated r times, the division would be $2^n \times r$. It is a simple algebraic exercise to justify this procedure for finding sums of squares. We shall do so for the case of a main effect only. Justification of the interaction sums of squares is left as an exercise to the reader.

For a three-factor analysis with only one observation per cell the sum of squares for the first main effect is

$$4 \times \Sigma_i (x_{i..} - x_{...})^2 \qquad (6)$$

in the notation of Chapter 3. Now, for the two permitted values of i we get

$$x_{i..} = \tfrac{1}{4}(a + ab + ac + abc) \text{ and } \tfrac{1}{4}((1) + b + c + bc)$$

Also

$$x_{...} = \tfrac{1}{8}(a + ab + ac + abc + (1) + b + c + bc)$$

On substituting in (6) the two terms to be summed become identical to

$$\frac{1}{64}(a + ab + ac + abc - (1) - b - c - bc)^2$$

Hence the sum of squares becomes

$$4 \times \frac{2}{64}(a + ab + ac + abc - (1) - b - c - bc)^2$$

which is $\tfrac{1}{8}$ of the square of the value given by (3). Thus the suggested procedure gives the correct sum of squares.

To illustrate the simplicity with which the analysis of variance table can be constructed in this special case, we shall use a simple example. The method of calculation is simplified by the use of a tabular method due to Yates (1937). The data for the eight experiments can be tabulated as follows:

Treatment	(1)	c	b	bc	a	ac	ab	abc
Result	11·8	20·9	8·5	16·2	9·9	18·3	8·1	16·0

The data are tabulated by including one variable at a time as shown and then differencing the data in pairs. Thus for X, the figures correspond to $(1) + a$, $b + ab$, $c + ac$, $bc + abc$, $a - (1)$, $ab - b$, $ac - c$, $abc - bc$. Columns Y and Z are obtained similarly. Column Z then corresponds to total effect, and hence the appropriate sum of squares can be calculated. If the design has been repeated, this method of calculation can be used on the totals for each combination of factors.

Treatment	Result	X	Y	Z	Effect	(Effects)²/8
(1)	11·8	21·7	38·3	+109·7	I	-
a	9·9	16·6	71·4	- 5·1	A	3·251
b	8·5	39·2	-2·3	- 12·1	B	18·301
ab	8·1	32·2	-2·8	+ 3·9	AB	1·901
c	20·9	-1·9	-5·1	+ 33·1	C	136·951
ac	18·3	-0·4	-7·0	- 0·5	AC	0·031
bc	16·2	-2·6	+1·5	- 1·9	BC	0·451
abc	16·0	-0·2	+2·4	+ 0·9	ABC	0·101

As only eight experiments have been carried out, we would not expect to learn much about the interactions. Hence we shall assume that they do not exist and pool the various sums of squares, and the final analysis of variance table becomes:

Source of variation	Sums of squares	Degrees of freedom	Mean squares
Factor A	3·25	1	3·25
Factor B	18·30	1	18·30
Factor C	136·95	1	136·95
Error	2·48	4	0·62
Total	160·98	7	

We conclude from this table that factors B and C are highly significant and there is also an indication that factor A may be affecting the results. If the interactions are to be investigated, it is desirable to repeat the whole experiment or increase the number of levels of the factors.

7.3 Confounding

When several factors are present, and particularly if the design is repeated several times, it is necessary to carry out a large number of experiments. It may happen that the experimental material is not sufficiently homogeneous to accomodate all the experiments. Let us suppose that the material occurs in homogeneous blocks which are large enough to hold half of the required experiments. There is now an effect of blocks to be allowed for in the analysis. It is possible to arrange the experiments over two blocks so that the block effect only occurs in the estimate of one of the interactions. Thus it is possible still to get tests of significance for main effects and most of the interactions, although the block effect may be significant. The block effect is said to be "confounded" with the interaction, and tests of significance for neither are available, hence the confounded interaction should be a high one. The allocation of the individual experiments between the blocks in order to achieve this state will now be described.

Consider a 2^3 analysis to be split over two blocks and it is desired to confound the ABC interaction with the block effect. Now

$$ABC = (a-1)(b-1)(c-1)$$
$$= (abc + a + b + c) - (ab + ac + bc + (1))$$

We shall assign the four experiments in the first bracket to one block and the remaining treatments to the second block. All the other effects are then independent of block differences, as shown by the following table. This shows the signs to be applied to each experiment when estimating the various effects.

Block	1				2			
Experiments	abc	a	b	c	ab	ac	bc	(1)
Effects ABC	+	+	+	+	−	−	−	−
A	+	+	−	−	+	+	−	−
B	+	−	+	−	+	−	+	−
C	+	−	−	+	−	+	+	−
AB	+	−	−	+	+	−	−	+
AC	+	−	+	−	−	+	−	+
BC	+	+	−	−	−	−	+	+

For all effects, except ABC, there are exactly equal numbers of positive and negative signs within each block. Hence these effects are independent of block effects.

As an example, consider the data of section 7.2 split between two blocks of raw material as described above. Further, suppose that the design is repeated seven times, using 14 blocks in all. The analysis of variance table would look as shown at the top of the next page.

Unless we are prepared to assume Model 1, tests of significance for the main effects are not available, as shown in Chapter 3. Assuming Model 1, the main effects and first-order interactions can be tested against the error mean square. If one is further prepared to assume that the second-order interaction is zero, then a test for block effect can also be made against the error mean square. The block effect and the ABC interaction are bound together in the first line of the table and no analysis will be able to separate them.

So far we have assumed that a block of material will carry half of the desired experiments. In fact, we have divided a 2^3 design into two blocks of four experiments each. It may be necessary to divide it into four blocks of two experiments. The block effect will now have three degrees of freedom, and therefore it will be necessary to select three

Source of variation	Sums of squares	Degrees of freedom	Mean squares
Blocks and ABC	9·86	13	0·76
A	112	1	112
B	26·71	1	26·71
C	364	1	364
AB	1·87	1	1·87
AC	13·11	1	13·11
BC	0·02	1	0·02
Error	10·43	36	0·29
Total	538·00	55	

of the interactions to confound with block differences. In practice, when any two have been selected arbitrarily, it is found that the third is uniquely determined. The third interaction is found by multiplying the first two together and replacing squared terms by unity.

For example, we shall confound the ABC and BC interactions in a 2^3 design. We shall find that we have also confounded a third effect, namely the main effect of factor A. To divide the experiment into four blocks we must first expand the two confounding interactions:

$$ABC = (abc + a + b + c) - (ab + ac + bc + (1))$$
$$BC = (abc + bc + a + (1)) - (ab + ac + b + c)$$

The four blocks can now be uniquely assigned by considering the combinations of signs in the two expressions above.

Block 1 with two + signs will contain abc and a.
Block 2 with + and − signs will contain b and c.
Block 3 with − and + signs will contain bc and (1).
Block 4 with two − signs will contain ab and ac.

With this configuration, it will be found from examination of the following table that all the remaining effects (except for factor A) are independent of any possible block effects. As before, the table shows the signs to be attached to each experiment when estimating the effects.

	Block	1		2		3		4	
	Experiments	abc	a	b	c	(1)	bc	ab	ac
	ABC	+	+	+	+	−	−	−	−
	BC	+	+	−	−	+	+	−	−
	A	+	+	−	−	−	−	+	+
Effects	B	+	−	+	−	−	+	+	−
	C	+	−	−	+	−	+	−	+
	AB	+	−	−	+	+	−	+	−
	AC	+	−	+	−	+	−	−	+

For the last four rows of the table, there are exactly equal numbers of positive and negative signs within each block. Hence these effects are independent of block effects. The three effects ABC, BC and A have been confounded with the unknown block effects.

It is not good practice to confound a main effect, and it would have been better above to confound the three first-order interactions AB, BC and AC since the primary object of designs of the type we are considering is to test the significance of main effects. In a 2^4 design in which the four factors are A, B, C and D, the best way of confounding into two blocks of eight experiments will be by confounding the $ABCD$ interaction. If it is necessary to confound into four blocks of four experiments, then it is usual to confound the interactions AB, CD and $ABCD$ or the interactions ABC, BCD and AD. Because the third effect is determined once the first two have been allocated, it will be found impossible to exclude both main effects and first-order interactions from the set of three which are to be confounded. The methods of confounding can be extended to as many factors as required. For example, five factors may be confounded into four blocks of eight or eight blocks of four. In the second case it will be found that seven of the possible interactions have been lost, corresponding to the seven degrees of freedom of the block effect.

7.4 Fractional replication

The total number of experiments doubles with every new factor to be examined in the design. When the number of factors exceeds five, this total can become excessive. For reasons of economy or for lack of experimental material, it may be impossible to carry out all these experiments. The problem then arises of doing only a fraction of the total and still getting worthwhile results. In exchange for making fewer experiments we would be prepared to sacrifice the higher-order interactions, since these are unlikely to be real and certainly of little interest.

As an example, let us consider a 2^4 experiment on which we propose to carry out eight experiments only. This is known as a "half replicated" design, since we are only carrying out half of the work required by a full factorial design. The experiments actually performed will correspond to one of the blocks obtained when the interaction $ABCD$ is confounded. These can be represented by (1), ab, ac, ad, bd, bc, cd, $abcd$. With only these eight results, the best estimate of factor A which we can get becomes $ab + ac + ad + abcd - (1) - bc - bd - cd$. Similarly, if we consider the estimate of BCD in terms of these results, it becomes $ab + ac + ad + abcd - (1) - bc - bd - cd$ which is the same as the result for A. This means that for a half-replicate, A is indistinguishable from BCD. A and BCD are said to be "aliases" and we write $A = BCD$. In the same way we find that AB is aliased with CD and, in general, every term has one alias which is given by multiplying that element by the term confounded (and replacing any squared terms by unity).

Thus for our eight observations which have 7 degrees of freedom, we may write

A	$=$	BCD	AB	$=$	CD
B	$=$	ACD	AC	$=$	BD
C	$=$	ABD	BC	$=$	AD
D	$=$	ABC			

If we had started by confounding ABC, we would have the seven aliases

A	$=$	BC	AD	$=$	BCD
B	$=$	AC	BD	$=$	ACD
C	$=$	AB	CD	$=$	ABD
D	$=$	$ABCD$			

The first set is the better since the aliases of the main effects are all second-order interactions and unlikely to be real.

Higher fractional replications can be built up on similar lines. The smallest quarter replicate that it is reasonable to use occurs with a 2^8 design. Suppose we have confounded $ABCDE$, $ABFGH$, $CDEFGH$, and the 2^6 experiments corresponding to only one block have been carried out. Each effect will now have three aliases which are formed by multiplying this effect by each of the confounding effects in turn. Thus we may write

$$A = BCDE = BFGH = ACDEFGH$$
$$AB = CDE = FGH = ABCDEFGH \quad \text{etc.}$$

Even when only a fraction of the total experiment is carried out the number of individual experiments may be large. For example, a half-replicate of a 2^8 experiment will involve $2^7 = 128$ separate runs. For such a design, the raw material may only be homogeneous over very small blocks, which would not accommodate 128 experiments. Consequently, there is a need for confounding to eliminate block differences in fractionally replicated experiments, just as there was for the fully replicated experiments of section 7.3.

The application of confounding follows in exactly the same way as before. Any interaction (together with its alias) may be confounded. It is not normal practice to choose the highest-order interaction for confounding, since the alias may be a main effect and this will also be confounded with block effects. In a half-replicate of a 2^6 experiment, we could confound into two blocks of 16 by selecting any of the second-order interactions, e.g. ABC and then its alias DEF will also be confounded. To confound into four blocks of eight, a set of three interactions must be selected as before, e.g. ABC, ADE, and $BCDE$. Then their aliases DEF, BCF and AF will be lost. It is impossible not to lose one first-order interaction in this latter case.

When computing the results of a fractionally replicated experiment, use may be made of Yates' method if care is taken to order the results correctly. For example, consider a half-replicate of a 2^4 experiment, which is formed by confounding the effect $ABCD$. The eight experiments can be represented by (1), ab, ac, ad, bd, bc, cd, $abcd$. If we temporarily ignore factor D, these become (1), ab, ac, a, b, bc, c, abc, which is a complete replicate of a 2^3 experiment. Hence the results may be ordered and analysed as in section 7.2. As before, the effects are identified as I, A, B, AB, C, AC, BC, ABC. Because of their aliases these are equivalent to I, A, B, AB, C, AC, BC, D. Therefore, if we are prepared to ignore second- and third-order interactions, an analysis of variance table can be drawn up to test the significance of the main effects. The three first-order interactions which have been estimated are usually not tested, since their aliases are also first-order interactions. A significant result would show that either the interaction or its alias or both was important.

7.5 Factors having three or four levels

Despite the great utility of 2^n designs, there are cases where it is practical to vary the factors over a fairly wide range, e.g. power-station boilers may be tested at light load on Sunday mornings. In these cases factors may be tested at three, four, or even more levels, and analysis of these experiments can be carried out comparatively easily by conventional analysis of variance. Let us suppose that we are investigating the process of machining steel shafts. We may try two cutting compounds, the factory standard, a_0, and a new one under test a_1. The experiment may be analysed for three different feed rates b_0, b_1, and b_2.

This would be described as a 3×2 design, and if there were 4 feed rates it would become a 4×2 design. If to the three feed rates we add a third variable, cutting speed, at 4 different levels, we have now to analyse a $4 \times 3 \times 2$ design.

When a factor has more than two levels, special forms of analysis have been derived for specific purposes. For example, in a 3^2 design it is possible to estimate the linear and quadratic effects of the two factors and also the interactions between these effects. For the main effects of factor A we have

$$
\begin{aligned}
A_L &= (a_2 b_2 + a_2 b_1 + a_2 b_0) - (a_0 b_2 + a_0 b_1 + a_0 b_0) \\
&= (a_2 - a_0)(b_2 + b_1 + b_0) \\
A_Q &= (a_2 b_2 + a_2 b_1 + a_2 b_0) - 2(a_1 b_2 + a_1 b_1 + a_1 b_0) \\
&\quad + (a_0 b_2 + a_0 b_1 + a_0 b_0) \\
&= (a_2 - 2a_1 + a_0)(b_2 + b_1 + b_0)
\end{aligned}
$$

and there are similar formulae for factor B. The interactions are estimated by

$$A_L B_L = (a_2 b_2 - a_0 b_2) - (a_2 b_0 - a_0 b_0)$$
$$= (a_2 - a_0)(b_2 - b_0)$$

$$A_L B_Q = (a_2 b_2 - a_0 b_2) - 2(a_2 b_1 - a_0 b_1) + (a_2 b_0 - a_0 b_0)$$
$$= (a_2 - a_0)(b_2 - 2b_1 + b_0)$$

$$A_Q B_L = (a_2 b_2 - 2a_1 b_2 + a_0 b_2) - (a_2 b_0 - 2a_1 b_0 + a_0 b_0)$$
$$= (a_2 - 2a_1 + a_0)(b_2 - b_0)$$

$$A_Q B_Q = (a_2 b_2 - 2a_1 b_2 + a_0 b_2) - 2(a_2 b_1 - 2a_1 b_1 + a_0 b_1)$$
$$+ (a_2 b_0 - 2a_1 b_0 + a_0 b_0)$$
$$= (a_2 - 2a_1 + a_0)(b_2 - 2b_1 + b_0)$$

To each of these four main effects and four interactions may be allocated one degree of freedom from eight available from the experiment. To build up an analysis of variance table to test for the significance of these effects, the sums of squares are obtained by squaring the values obtained above and dividing by the appropriate constant. The divisors are easily obtained from the symbolic equations given above. In each bracket, the coefficients of the a's and b's are squared and summed. Then the two numbers obtained from the two brackets are multiplied together to give the divisor. For a discussion of this and other special forms of analysis the reader should examine Bennett and Franklin (1954).

Term	A_L	A_Q	B_L	B_Q	$A_L B_L$	$A_L B_Q$	$A_Q B_L$	$A_Q B_Q$
Divisor	6	18	6	18	4	12	12	36

CHAPTER 8

FURTHER DISCUSSION

8.1 Introduction

The object of this chapter is to discuss points which arise only occasionally in analysis of variation applications. In particular, we shall cover in more detail the types of models which may be assumed for the data and secondly, we shall indicate how confidence limits may be applied to linear combinations of variances occurring in this work.

8.2 Models

Throughout this monograph we have employed such terms as Model I, Model II and Mixed Model, depending on the nature of the factors in the experiment. For Model I it was assumed that for all the factors the levels used in the experiment constitute the population of levels in which we are interested. When the levels used form a sample from an infinite population of levels, Model II is appropriate. Tukey (1949) emphasised the restrictiveness of these models and proposed to extend the range by defining the following:

Model III. This is the same as Model I except that the error term is now assumed to be sampled from a population the size of which is equal to the number of determinations in the experiment.

Model IV. The levels for the factors and the error term are samples from finite populations. This includes Model III as a special case.

Model V. The individual effects and interactions and error terms are random samples of one from infinite populations. This includes Model II as a special case.

Model X is a very general model of little practical utility. In this, all elements of the original linear equation are assumed to be the sum of three independent sub-elements, the first satisfying Model III, the second Model IV and the third Model V.

The most general of these models still does not allow for non-independence between elements of the linear equation which are not of the same kind. Apart from Models I and II, the more sophisticated models mentioned above have received little attention in statistical literature, except in the more mathematical journals. An excellent review of the present state of the use of models is given by Plackett (1960).

The method adopted by Bennett and Franklin (1954) is the special case of Model IV, in which the error term is assumed to be a random variable from a Normal distribution. The levels of the factors are assumed to be random samples from a finite population of levels, and use is made of the fact that the variance of the mean of a random sample of p from a finite population of size P with variance σ^2 is given by $\left(1 - \frac{p}{P}\right)\frac{\sigma^2}{p}$. In this way a table of the expected values of the mean squares for a two-way crossed classification as described in section 2.2 can be built up. Suppose that the p levels of factor A have been chosen from a population of size P, and the q levels of factor B come from a population of size Q, then the expected values of the mean squares become

$$nq\sigma_A^2 + n(1 - \frac{q}{Q})\sigma_I^2 + \sigma^2$$

$$np\sigma_B^2 + n(1 - \frac{p}{P})\sigma_I^2 + \sigma^2$$

$$n\sigma_I^2 + \sigma^2$$

$$\sigma^2$$

From this table, it is seen that a significance test is only available to test for existence of interaction in the general case. To develop tests for the main effects we must limit ourselves to certain special cases. When the sample of levels corresponds to the whole population we can put $p = P$ and $q = Q$; we get tests for the main effects by dividing the mean squares by the "within cells" mean square. This corresponds to Model I of our previous analysis. In the case when the size of the population of levels becomes infinite, then the main effects are tested against the interaction mean square. This corresponds to Model II of our previous analysis. Our Mixed Model can also be directly deduced from the above table by putting $P = \infty$ and $q = Q$, since we assumed that factor A was to have a random effect. If the nature of the two factors is interchanged, then we must make the appropriate alterations in the above table.

8.3 Linear combinations of variances

Expressions which can be expressed as linear combinations of variances occur frequently in analysis of variance. Two of the most frequent examples will be described here.

(i) *Gross variability*. This is the name given to the variance of a single individual obtained at random under Model II. It may also be described as the variance of the grand population from which the individuals have been selected and can be expressed as a linear function of the associated mean squares.

One of its main applications is in industrial problems where, in testing a new process, a sample is taken from a load of raw materials, and from each unit of the sample a number of articles are manufactured. The articles will vary, due to lack of homogeneity in the sample and also due to differences in the manufacturing. From the consumer's point of view it is the total variation of the individual articles which is important, rather than the manufacturer's variation.

In the Mixed Model, gross variability corresponds to the variance of an individual obtained at random for given levels of the non-random factors.

(ii) *Components of variance*. In Model II and the Mixed Model we introduced variances corresponding to the random effects and the interaction. These have been termed "components of variance" since they represent the parts of the total variation attributable to these factors and interactions. It is not difficult to see from any of the cases considered in this test that these components can be expressed as linear combinations of the associated mean squares.

Variance components have been used in studying the precision of instruments and are useful, in general, in ascertaining the factor which contributes most to the variation of the individual, so that action can be taken to decrease the effect of this factor. Another use has been found by Cameron (1951) who has considered the use of variance components in determining the precision of estimating the clean content of wool.

The problem of deriving confidence limits for linear combinations of the mean squares of the types described above is one which offers no neat mathematical solution. The first attempt to deal with this problem was made by Fisher (1935) in a paper on fiducial probability. The only further investigation along the lines suggested by Fisher was made by Bross (1950) who gives a method of finding approximate fiducial limits for variance components.

Working along the lines of symbolic Taylor expansions, Huitson

(1955) and Welch (1956) have derived the leading terms of a series expansion for confidence limits to such expressions and Huitson (1955, 1958) has given tables of this expansion. Expansions of this kind have been thoroughly investigated and behave satisfactorily, except perhaps for small degrees of freedom. Reference may also be made to a solution proposed by Bulmer (1957).

8.4 Suggestions for further reading

Apart from the works which have been given specific mention in the text, the following will be found useful for further reading.

Brownlee (1951). This is a non-mathematical text which describes the arithmetical steps in carrying out an analysis, including analyses of confounded experiments and fractional replications.

Fisher (1960). A profound discourse on the aims and objects of experimentation. Not a book for the beginner, but will repay study.

Kempthorne (1952). This gives a more comprehensive approach to the subject than has been attempted in this monograph.

Scheffé (1959). A book containing a full description of the subject but aimed at the pure mathematician rather than the practising statistician.

Kendall and Stuart (1958, 1961). Among many other branches of statistics there is much of value on the subject of analysis of variance. This work must be included in any list, since it is the standard reference book for theory.

REFERENCES

Bartlett, M.S. (1947). The use of transformations, *Biometrics*, **3**, 39.

Bennett, C.A. and Franklin, N.L. (1954). *Statistical Analysis in Chemistry and the Chemical Industry*, New York: John Wiley & Sons; London: Chapman & Hall Ltd.

Bross, I. (1950). Fiducial intervals for various components, *Biometrics*, **6**, 136.

Brownlee, K.A. (1951). *Industrial Experimentation*, H.M.S.O.

Bulmer, M.G. (1957). Approximate confidence limits for components of variance, *Biometrika*, **44**, 159.

Cameron, J.M. (1951). Use of components of variance in preparing schedules for the sampling of baled wool, *Biometrics*, **7**, 83.

Cochran, W.G. and Cox, G.M. (1957). *Experimental Designs*, New York: John Wiley & Sons; London: Chapman & Hall Ltd.

Crump, S.L. (1946). The estimation of variance components in analysis of variance, *Biometrics*, **2**, 7.

Davies, O.L. (1960). *Design and Analysis of Industrial Experiments*, Edinburgh: Oliver & Boyd.

Eisenhart, C. (1947). The assumptions underlying the analysis of variance, *Biometrics*, **7**, 1.

Fisher, R.A. (1935). The fiducial argument in statistical inference, *Annals of Eugenics*, **6**, 391.

———— (1954). *Statistical Methods for Research Workers*, 12th edn, Edinburgh: Oliver & Boyd.

———— (1960). *The Design of Experiments*, 7th edn, Edinburgh: Oliver & Boyd.

Huitson, A. (1955). A method of assigning confidence limits to linear combinations of variances, *Biometrika*, **42**, 471.

———— (1958). Further critical values for the sum of two variances, *Biometrika*, **45**, 279.

Kempthorne, O. (1952). *The Design and Analysis of Experiments*, New York: John Wiley & Sons.

Kendall, M.G. and Stuart, A. (1958, 1961). *The Advanced Theory of Statistics*, London: Griffin & Co.

Plackett, R.L. (1960). Models in the analysis of variance, *J. R. Statist. Soc.*, B, Part 2.

Satterthwaite, F.E. (1946). An approximate distribution of estimates of variance components, *Biometrics*, **2**, 110.

Scheffé, (1959). *The Analysis of Variance*, New York: John Wiley & Sons.

Tukey, J.W. (1949). Dyadicanova, an analysis of variance for vectors, *Human Biology*, **21**, 65.

Welch, B.L. (1936). Specification of rules for rejecting too variable a product, with particular reference to an electric lamp problem, *Suppl. J. R. Statist. Soc.*, **3**, 29.

─────── (1956). On linear combinations of several variances, *J. Amer. Statist. Assoc.*, **51**, 132.

Wernimont, T.G. (1947). Quality control in the chemical industry, *Industrial Quality Control*, **3**, 5.

Wishart, J. and Saunders, H.G. (1936). *Principles and Practice of Field Experimentation*, Empire Cotton Growing Corporation.

Yates, F. (1937). The design and analysis of factorial experiments, Imperial Bureau of Soil Science, Technical Communication 35.

INDEX

Aliases, 72, 74
Analysis of variance table, 4, 14, 17
Analysis of covariance, 56-63

Balanced incomplete blocks, 48
BARTLETT, M.S., 36
BENNETT, C.A. & FRANKLIN, N.L., 3, 75, 77
BROSS, I., 78
BULMER, M.G., 79

CAMERON, J.M., 78
COCHRAN, W.G. & COX, G.M., 21, 51
Components of variance, 78
Confidence limits, 76, 78
Confounding, 68, 73
Covariance, 56
Crossed classification, 11
CRUMP, S.L., 22

DAVIES, O.L., 48

EISENHART, C., 3
Experimental design, 43

F-distribution, 5
Factors, 1, 2
Finite populations, 77
FISHER, R.A., 1, 78
Fractional replication, 72

Graeco-Latin square, 48
Gross variability, 78

Half replicate, 72
HUITSON, A., 78, 79
Hypothesis testing, 5

Interaction, 11, 15

Latin cube, 48
Latin square, 46
Levels, 2, 3

Linear regression, 37
Low F-ratio, 7

Mean squares, 4
Missing data, 21
Models, 3, 76

Nested classification, 11, 13, 19
Nested and crossed analyses, 32
Non-significant interaction, 15
Normal distribution, 4

One-way classification, 3-8
Overall mean, 3, 8

PLACKETT, R.L., 3, 77
Poisson distribution, 36

Random factors, 3
Randomisation, 2
Randomised blocks, 44
Regression analysis, 37, 58

SATTERTHWAITE, F.E., 29
Small numbers of levels, 64
Split plots, 49
Systematic factors, 2

Transformations, 35, 36
TUKEY, J.W., 76
Two-level experiments, 64-75
Two-way classification, 11-21

Unequal numbers of observations, 6

Variance, 1, 4

WELCH, B.L., 29, 79
WERNIMONT, T.G., 24
WISHART, J. & SAUNDERS, H.G., 56

YATES, F., 67
Yates method, 67, 73